DRUGS, RITUALS AND ALTERED STATES OF CONSCIOUSNESS

DRUGS, RITUALS AND ALTERED STATES OF CONSCIOUSNESS

edited by

BRIAN M. DU TOIT
University of Florida

A. A. BALKEMA / ROTTERDAM / 1977

ISBN 90 6191 014 5

Printed in the Netherlands

To Eleanor Carroll
who has meant much
to many of us

CONTENTS

PREFACE

This volume is an outgrowth of a symposium held during the 1975 annual meetings of the Society for Applied Anthropology in Amsterdam, the Netherlands. Prior to the meetings we decided to explore the realm of drug use and its relationship to secular ritual. Papers were presented inter alia by Agar, Carter, du Toit, and Partridge while Jan van Baal, at the time professor of Anthropology at Utrecht University acted as discussant.

These original papers have been expanded and a number of additional invited papers added. To each of these contributors I would like to express my sincere appreciation.

A word of special thanks must be offered to Mr A. T. Balkema. After the Amsterdam conference he immediately expressed his willingness to publish this volume, and in the intervening months he not only showed exceptional patience but also gave this publication his personal care.

BRIAN M. DU TOIT

INTRODUCTION

Somewhat belatedly the social sciences are entering the realm of drug studies. This does not mean that earlier observers did not remark on the presence of various mind altering substances, but now for the first time social scientists are giving their attention to the normal ethnographic setting and use of such products. While alcohol and opiate use did receive some attention in the past, the study of psychoactive and hallucinogenic drug use has become acceptable within the past decade. Even the other sciences are increasingly recognizing the need to supplement their epidemiological research with studies of drugs in the normal community setting.

A number of early writers did give their attention to such mind altering substances. Opler (1970: 40-41) reproduces an early report of how the settlers discovered the properties of the jimson weed.

In 1676 a rather hungry contingent of British redcoats arrived in Jamestown, Virginia, to quell an uprising known as Bacon's Rebellion. While bivouacked there, the soldiers gathered some young plants and cooked themselves a tasty potherb. They called the plants the James Town weed; botanists classify the species as *Datura stramonium.* The consequences of this historic meal are to be found in Robert Beverly's 'History and present state of Virginia':

> The James-town Weed (which resembles the Thorny Apple of Peru, and I take to be the Plant so call'd) is supposed to be one of the greatest Coolers in the World. This being an early Plant was gather'd very young for a boil'd salad, by some of the Soldiers sent thither, to pacifie the Troubles of Bacon; and some of them eat plentifully of it, the Effect of which was a very pleasant Comedy; for they turn'd natural Fools upon it for several Days. One would blow up a Feather in the Air: another would dart Straws at it with much Fury; and another stark naked was sitting in a Corner, like a Monkey, grinning and making Mows at them; a Fourth would fondly kiss, and paw his Companions and snear in their Faces, with a Countenance more antick, than any in a Dutch Droll. In his frantick

Condition they were confined, lest they should in their Folly destroy themselves; though it was observed that all their Actions were full of Innocence and good Nature. Indeed, they were not very cleanly; for they would have wallow'd in their own Excrements, if they had not been prevented. A Thousand such simple Tricks they play'd, and after Eleven Days, return's themselves again, not remembering any thing that had pass'd.

This was of course the same narcotic used by Native Americans to alter their states of consciousness and produce visions. Perhaps as well known was the fly agaric, *Amanita muscaria,* a psychoactive mushroom. Its use among the shamans of Siberian tribes was described by Jochelson around the turn of the century as producing "intoxication, hallucinations, and delirium". These were particularly useful during seances (1905-08: 583). There are strong suggestions that the ambrosia imbibed by the mythological heroes of classical Greece also was a decoction of *Amanita muscaria* (Furst 1972: X).

One of the most thorough surveys of the previous century was the Report of the Indian Hemp Drug Commission, 1893-94. It recorded the hallucinogenic effects of cannabis in India and the veneration given by many to the plant. But such treatment of plants is certainly not restricted to the cannabis plant. In her study of the Zapoteco speakers in Oaxaca, Mexico, Parsons records the drinking of an infusion of leaves, and the eating of certain seeds which caused trancelike sleep and visual hallucinations. This clematis-like vine and other plants "are 'spoken to', that is, prayed to. It is almost certain that marigold and other ritually or medicinally used plants were once thought of in similar animistic terms" (Parsons, 1936: 228). The value attached to these various plants was a direct outflow of experiences related to sensory stimulations and altered moods.

The first fulfledged study of the socio-ritual uses of an hallucinogen was La Barre's study (1938) of peyote, *Lophophora williamsii.* But it has really been within the last decade that many social scientists have forsaken the traditional ethnological field and turned increasingly to focus on drug use and drug abuse. This interest in part flows from the need to understand the use of mind and mood altering substances in the community setting, to view them as a part of community life, and to fathom the importance of these various substances in ritual, social, and interpersonal contexts. But we should never loose sight of the important fact, as Sorenson in this volume points out, that the concept "mind altering" is in part culturally defined. What is more, it may also be situationally defined. A matter which is very closely related to this definition is that of social acceptability and cultural patterning. It is well known that cannabis use is accepted much more readily in East Indian and African communities than in Euro-American communities. In addition to being acceptable there are long standing patterns of use which show up in studies of first drug use and justifications for use. Harding and Zinberg, in this volume contrast the misuse of alcohol by American Indians while they have used jimson weed and peyote in controlled forms. Also in this volume du Toit shows how in the multi-ethnic South African situation the African and Indian ethnic groups have a clear historical pattern of cannabis as the most important substance used. But the research reports presented here do not only show the traditional emphasis.

Along with the broadening of focus on the part of research workers has also come a shift away from necessarily studying other societies and an increasing concentration on their own modern urban society. In part this shift reflects the greater acceptability of anthropology and the degree to which anthropologists, along with other social scientists, have something meaningful and important to contribute. Thus research funding and employment possibilities place these persons now not only in traditional communities and culturally heterogeneous villages but also among their own ethnic group on street corners, in bars, and among street gang members where they observe, interact with, and study the users of various kinds of drugs.

These research situations, be they the traditional anthropological field study or the modern study of drug sub-cultures, present their own problems. Mercer & Smart (1974: 308) delineate three major problem areas in such epidemiological studies:

> (1) particularly for primitive populations, there are few competent researchers willing to go into the area and investigate; (2) the standard technique of questionnaires is of little use in an illiterate population; and (3) obtaining reliable information from a group that does not want to be investigated is a particularly difficult task.

It is of course in the first two areas that anthropologists have made their major contributions, e.g. Marlene Dobkin de Rios, Peter Furst, Michael Harner, Barbara Myerhoff, Gerardo Reichel-Dolmatoff, and Johannes Wilbert, to mention only a few. In the latter of the three problem areas mentioned they are currently employing their methods of direct observation, small group analysis, networks and similar approaches. For a valuable collection of studies along these lines, though limited to studies of cannabis, the reader is referred to Rubin (1975). Even greater time depth is to be found in the field of alcohol studies — where Heath (1975) has presented a very thorough summary. It is indeed fortunate that social scientists have received such wholehearted support from specialists in related fields. Best known and most distinguished are perhaps the botanist Richard Evans Schultes, the biologist William A. Emboden, and the amateur ethnomycologist R. Gordon Wasson to mention only three.

Along with actual field studies of mind altering substances, has come a growing attempt at understanding exactly what is involved in this process. What meaning does it have, and to what extent are we dealing with new forms of cultural patterning, new regularities, new rituals. But we are frequently reminded that altered states of consciousness need not be drug induced (Bourguignon, 1973), that hallucinations are extremely common both in normal dreaming and thinking (Hartmann, 1975), as well as in religious fervor (Goodman, 1974; Henney 1974), even though religious ecstasy may be drug induced (Fernandez, 1972).

It was a combination of these various interests which lead to the compilation of this volume. We hope it raises as many questions as it answers and that international scholarly communication will flow from its publication.

BIBLIOGRAPHY

Bourguignon, Erika (ed.), 1973, Religion, altered states of consciousness and social change. Columbus, Ohio: State University Press.

Fernandez, James W., 1972, Tabernanthe Iboga: narcotic ecstasis and the work of the ancestors. In Peter T. Furst (ed.), Flesh of the Gods. New York: Praeger Publishers.

Furst, Peter T., 1972, Introduction. In Peter T. Furst (ed.), Flesh of the Gods. New York: Praeger Publishers.

Goodman, Felicitas D., 1974, Disturbances in the Apostolic Church: A trance-based upheaval in Yucatán. In Felicitas D. Goodman et al (eds.), Trance, healing, and hallucination. New York: John Wiley & Sons.

Hartmann, Ernest, 1975, Dreams and other hallucinations: an approach to the underlying mechanism. In R. K. Siegel and L. J. West (eds.), Hallucinations: behavior, experience and theory. New York: John Wiley & Sons.

Heath, Dwight B., 1975, A critical review of ethnographic studies of alcohol use. In Robert J. Gibbins et al. (eds.), Research advances in alcohol and drug problems, Vol. 2. New York: John Wiley & Sons.

Henney, Jeannette H., 1974, Spirit-possession belief and trance behavior in two fundamentalist groups in St. Vincent. In Felicitas D. Goodman et al. (eds.), Trance, healing and hallucination. New York: John Wiley & Sons.

Jochelson, Waldemar I., 1905-08, The Koryak. Report of the Jessup Expedition, 1900-1901. Memoirs of the American Museum of National History, Vol. 10, part 2.

La Barre, Weston, 1938, The peyote cult. New Haven: Yale University Publications in Anthropology No. 19.

Mercer, G. W. & R. G. Smart, 1974, The epidemiology of psychoactive and hallucinogenic drug use. In Robert J. Gibbins et al. (eds.), Research advances in alcohol and drug problems, Vol. 1. New York: John Wiley & Sons.

Opler, Marvin K., 1970, Cross-cultural uses of psychoactive drugs (ethnopsychopharmacology). In W. G. Clark & J. del Giudice (eds.), Principles of psychopharmacology. New York: Academic Press.

Parsons, Elsie Clews, 1936, Mitla, town of the souls. Chicago: University of Chicago Press.

Rubin, Vera (ed.), 1975, Cannabis and culture. The Hague: Mouton Publishers.

THE SUBJECT AND THE FIELD

In using the term "drug" as part of our title, we are really attempting the widest coverage. A drug should be seen as any chemical substance, natural or artificial, which alters perception, mood, or consciousness. This would include natural products such as cannabis as well as productions such as alcohol, methaqualone and heroin (all of which are discussed later), although it would seem that only heroin does not produce a psychic effect thus altering the state of consciousness.

The two chapters in this first section introduce the topic which forms the subject of this book. In the first Bourguignon discusses altered states of consciousness which may be brought about with or without the aid of hallucinogens and other psychotropic drugs. While not necessarily elaborated in later chapters, this subject is inherent in most of the studies.

In lieu of a chapter specially prepared to discuss the botanical and chemical nature of the drugs covered by this book, we are extremely grateful to the author and the Editor-in-Chief for permission to reprint Schultes' study of hallucinogens. Since it was originally published in a review not normally read by social scientists and since our knowledge about these matters has not greatly advanced in the intervening five years, this chapter presents a current view. It does of course concentrate on hallucinogens, and in this sense ties in well with the final chapter.

ERIKA BOURGUIGNON

1

ALTERED STATES OF CONSCIOUSNESS, MYTHS, AND RITUALS

Altered states of consciousness (ASC) are universal human phenomena, which, like other such universals, are subject to a great deal of cultural patterning, stylization, ritualization, and rationalizing mythology. The concept of ASC is a broad, high-level abstraction covering such a variety of states as the various phases of sleep and dreaming, drunkenness, meditative states, and visionary states, possession trance (i.e., so-called "possession states"), as well as fever delirium, somnambulance, and many more. A. M. Ludwig (1972 (orig. 1966):11) defines them as

> any mental state(s), induced by various physiological, psychological, or pharmacological maneuvers or agents, which can be recognized subjectively by the individual himself (or by an objective observer of the individual) as representing a sufficient deviation in subjective experience or psychological functioning from certain general norms for the individual during alert, waking consciousness.

He lists a total of some 62 categories of states in five groups (Ludwig, 1968). The five groups are: reduction of exteroceptive stimulation and/or motor activity; increased alertness or mental involvement; decrease of alertness or relaxation of critical faculties, and, finally, presence of somatopsychological factors with states resulting primarily from changes in body chemistry, whether spontaneous (as in hormonal disturbances) or induced (as through the administration of pharmacological agents) (Ludwig, 1968: 71-75). In spite of the many different types of states and the varying manner of induction, Ludwig finds that they have a series of common characteristics such as alterations of thinking, disturbed time sense, loss of control, change in emotional expression, body image change, perceptual distortions, change in meaning or significance, a sense of the ineffable, feeling of rejuvenation and, lastly, hypersuggestibility. With regard to their function, he concludes that, "many altered states of consciousness serve as 'final common pathways' for many different forms of human expression, both adaptive and maladaptive" (Ludwig, 1972: 24).

Another classification of states of consciousness is presented by Roland Fischer, who has developed and refined what he terms a "cartography of inner space" (Fischer, 1971, 1974). He, too, begins with a normal state of routine waking consciousness, which represents a mid-point in a continuum of subcortical arousal. Higher arousal ranges from sensitivity, creativity, and anxiety to hyperaroused acute hyperphrenetic states to an ecstatic state of mystical rapture. Moving to the other side of the continuum, it ranges from normal relaxation to reduced arousal or tranquilization to the meditative state of zazen and on to hyporoused state of samádhi. This continuum is drawn in the form of a circle, with the normal waking state representing "the I" and the point at which both the highest aroused state of ecstatic mystical rapture and the most highly tranquilized state of samádhi meet defined as "the self". Fischer has added an evolutionary perspective to this cartographic description, arguing that humanity has moved from magical to mythical and then to mental structures of consciousness. This perspective is placed within the framework of a larger cosmology.

Anthropologists may wish to take issue with such an evolutionary view, if only because of the difficulty of studying states of consciousness over the millenia — or should one say the millions of years — of human evolution. We seem to have considerable difficulty in studying variation in states of consciousness among living peoples. None the less, a question of importance is raised by both Ludwig and Fischer: can we assume a "normal routine state of waking consciousness" as a given, essentially the same everywhere, even if we limit ourselves to contemporary groups of our species? We do know that cultural differences make for differences in perception (Segall, Campbell & Herskovits, 1966; Price-Williams, 1970); we know that groups with different cultures live in different behavioral environments (Hallowell, 1955, 1958). Indeed, such recent developments as ethnoscience and ethnomethodology have stressed the relativity of the "reality" within which people operate and which clearly must enter into the definition of the "normal waking state", in contrast to states of hyperarousal or hypoarousal. How similar, or how different, are the "normal" states? As we know, prehistoric and so called "primitive" contemporary peoples have shown remarkable ingenuity and resourcefulness in dealing with hostile environments, solving technological problems and accumulating funds of basic empirical knowledge on which we still draw. After all, the remarkable paintings of the Magdalenians, though they show us the magical and mythical mind at work, also show us solutions to difficult technological and technical problems. Answers to the questions about the normal state must come ultimately from psychobiology and biochemistry.

For the anthropologist, altered states are of particular interest for several reasons:

1. They represent an excellent opportunity (so far not widely exploited) for cooperative work between the various areas of anthropology:

psychological anthropology, study of ritual and comparative religion, ethnopharmacology, medical anthropology, physical anthropology with specific attention to the physiology of altered states, etc.

2. Somewhat more narrowly, they represent an opportunity for the investigation and comparison of the diverse ways in which basic biological and psychological givens are variously dealt with on the cultural plane: ranging from minimal to maximal stylization, from negative to neutral to positive evaluation, with rationalization or interpretation ranging from minimal interest to major mythological structures, etc.

I suggested earlier that altered states are universal human phenomena. Indeed, if we consider sleep and specifically the several phases of sleep (including REM sleep or dreaming) we must admit that all mammals, at least, pass through several states of consciousness. Human cultures differ in their stylization of sleep and they also differ in their stylization and evaluation of dreams. (See Bourguignon, 1972, for a review of the literature on dreams.) They similarly differ in their dealings with other states, both with regard to the amount of attention they pay to them, the affective evaluation of the states, the degree of ritualization, their integration into sacred beliefs, the mythological interpretations, etc.

In a five year, cross-cultural study of ASCs based on the anthropological literature (Bourguignon, 1968, 1973)* we found that in traditional societies at least some altered states are generally integrated into the systems of sacred beliefs and into dealings with supernatural or superhuman agencies (beings, powers). We found it meaningful to divide the behaviors, which range along a continuum, for analytic purposes according to the rationalizing ideology into two groups: states interpreted as due to possession by spirits, which we termed possession trance (PT) and states given some other interpretation, termed trance (T). We drew a world-wide sample of 488 societies from the 'Ethnographic Atlas' (Murdock, 1967) and found that ASCs were institutionalized within a religious framework in 90 per cent of these societies. In addition to these ASCs in the context of religious institutions, there was also in many societies a belief in a type of possession which did not manifest itself in an altered state; these beliefs, however, will not further concern us here. We also encountered, of course, many types of ASCs not institutionalized in the manner defined, and we did not include these in our codings. An example may be cited to clarify this distinction: E. E. Evans-Pritchard (1937:178) describes Azande "witch-doctors" who take "medicines which give them power to see the unseen and to enable them to resist great fatigue". We do not know whether these substances are in fact pharmacologically active, and Evans-Pritchard did not think so. The medicines are taken in conjunction with drumming and singing

* This study was conducted during 1963-68 and was supported in full by PHS grant MH 04763 from the National Institute of Mental Health.

and active dancing and the witch doctors, we are told, achieve a state of dissociation in which they prophesy, specifically identifying witches. This, then, is coded by us as T: there is an altered state, it occurs in a religious context, it is not explained as due to possession by some other entity or self. It may, or may not, be drug induced. Most witch doctors are men. According to a reference in another context (Evans-Pritchard, 1962) we know that the Azande also had a pattern of ghost-divination, in which women specialists became possessed by ghosts. Mention is also made of their eating special medicines, known only to them, but these appear to be unrelated to their possession trance. The Azande, as described by Evans-Pritchard thus represent a society having both T and PT, although we know a great deal more about the former than about the latter. They may, or may not, have used pharmacologically active substances to induce one of these states. They also used other means such as rhythmic sounds and motions, physical exertion, including dancing of a whirling type, leading to a degree of disorientation, possibly also involving hyperventilation, suggestion, and concentration. In addition to these ASCs involved in religious ritual and divination, there is also an indirect reference to the smoking of hemp by the Azande, for we are told that "no one may smoke hemp in a hut which lodges the oracle poision" (Evans-Pritchard, 1937:288). The reference here is to "benge" poison, used on chickens. Thus we may assume that the Azande know (or knew) a third, secular drug-induced type of ASC. It touches the magico-religious domain only through the taboo on contact between the secular activity of hemp smoking and the substance used in the poison oracle, which must be protected from various types of pollution. The occurrence of such practices, however, was not coded by us. Furthermore, among the disease entities recognized and named by the Azande, Evans-Pritchard lists epilepsy. This involves a type of ASC; however, it is not explained by spirit possession or, apparently, by other supernatural causes nor is there a report of a magical or religious curing ritual. The disease, in other words, is not integrated into a religious belief system, either with respect to explanation or treatment. Again, the occurrence of such states was not coded by us. Thus, our codings refer to ASCs institutionalized within a religious scheme, but omit at least two other kinds of states: spontaneous (perhaps generally pathological) states treated within an empirical or secular framework and induced states (perhaps generally drug-induced states) which are also outside a religious framework. It is clear that either or both types may be considered to be "ritualized". We shall return to this point presently.

It was mentioned earlier that we found sacred ASCs institutionalized in 90 per cent of our sample societies. If we erred, or if the literature on which our work was based erred, the error was surely in under-estimating the frequency of these patterns rather than in over-estimating it. There is, however, some variation in the incidence in the six major ethnographic regions of the world, and thus, we found altered states institutionalized in North America in 97 per cent of the societies, in South America in 85 per cent, in Sub-Saharan Africa in 82 per cent, in 80 per cent of Circum-Mediterranean societies, and in 94 per cent of the societies of both East Eurasia and the Insular Pacific.

Interestingly enough, the distribution of types of altered states, that is the presence of possession trance, trance or a combination of both, differs greatly for the various regions as well. If we contrast the Americas and Sub-Saharan Africa, specifically, we find that trance (T) alone is present in only 17 per cent of the African societies, but in 72 per cent of the North American societies and in 54 per cent of the South American ones. Possession trance (PT) alone is present in 45 per cent of the African sample of societies, but in only 4 per cent of the North American sample and in 8 per cent of the South American one. In 20 per cent of the African societies, furthermore, we find both trance and possession trance (T/PT) and this is true of 21 per cent of the North American and 22 per cent of the South American societies as well. Thus, while the predominant form of altered state in Africa is possession trance, in the Americas it is trance. Let us look more closely at the interesting contrasts between the New World and Sub-Saharan Africa. These regions are of particular interest since we are dealing here primarily with tribal peoples, with minimal intrusion of Old World Great Traditions and world religions.

North and South America

As has been indicated in the figures cited above, each major world area has its own characteristic "profile" with regard to the distribution of patterns of ASCs. Thus, we have seen that possession trance (PT) is relatively rare in the Americas and that the overwhelming number of societies in the Northern and Southern continent have trance (T). In North America, this is often linked to the widespread guardian spirit complex, with its vision quest. The vision is typically sought alone, and isolation may well be a factor in inducing the altered state involved here, as are fasting and perhaps some other forms of mortification, as well as the use of such drugs as datura and more recently peyote. Although in many tribes the vision quest is and was open to women as well as men, the general impression one gains from the ethnographic literature is that more men than women participated. The vision quest among American Indians was an individual and not a group activity. It established a private relationship, in many cases one not to be spoken of or to be revealed to others, between the supplicant and his guardian spirit. It gave him certain powers and he remembered the experience as the big event of his life. (This is in interesting contrast to the possession trancer who is generally amnesic for the events of the period of altered consciousness.)

Swanson (1973) has studied the typically North American guardian spirit quest as a rite of empowerment. The vision, through which the guardian is contacted, is one of the defining features of this complex. Swanson explicitly distinguishes it from possession and characterizes it as bestowing power to the individual as a gift that the beneficiary "on his own volition, might use or neglect" (Swanson, 1973:360). And, furthermore, "power was authenticated through its successful application" (p. 361). The seeker for a guardian spirit inducts himself, as it were, into adulthood. By contrast those who undergo corporate initiations, as in the widespread African puberty rites, are inducted by a group. Swanson sees the guardian spirit complex as characterized by individual initiative, and by the individual's own need and desire for the power to be achieved. He hypothesizes that such quest for a guardian spirit constitutes a test of the individual's independence and discretion and of the strength of his motives, and that these are important in societies where, in addition to collective interests, individuals have a large degree of autonomy in the pursuit of goals. Yet there must be collective interests as well, and the pursuit of individualism must not be extreme. Swanson takes his cue from the findings of Barry, Child & Bacon (1959) of a relationship between hunting and fishing subsistence economies and pressures toward self-reliance and achievement in socialization. Such societies should, in Swanson's view, however, not produce totally individualistic types, but persons concerned with the social order as well. He expects, therefore, and indeed finds, that the guardian spirit complex, entailing charisma of the office of adulthood, is present in societies emphasizing individualism, but not having such exclusive dependence on hunting and fishing as to produce total self-reliance and independence. Another expectation that is confirmed statistically is a rule of virilocal residence. And, finally, societies allowing in their political organization both for common and special interests are found to be more likely to have a guardian spirit complex than those with other types of decision-making processes. This study, which is based on a sample of 42 North American societies, concerns itself with the guardian spirit complex of which the vision is one diagnostic item. Consequently, it covers a smaller scope than our concern with the more varied concept of "trance".

In more recent religious movements among American Indians we find a change from stress on individual activity to an emphasis on group activities, as in the Ghost Dance or the peyote cult. But there is still concern for a private, inward passive experience. Drugs were often used in North America to induce trance. In a trial coding of 10 North American societies (Bourguignon, 1968) we identified 23 trance institutions, i.e. societally differentiated ritual situations in which an altered state of consciousness is induced. Of these, all but one involved trance rather than possession trance. We sought to identify sex of participants but in six cases this was not specified. We also sought to distinguish trance involving a group situation from trance involving only a single participant. We found no trance institutions involving women only; the single case of a trance institution involving mostly women concerned women in a group situation. Six institutions with both men and women trancers were divided equally between group and individual situations. Of the nine trance institutions involving all or mostly men, however, there were twice as many individual situations as group situations: only three group situations as compared to six situations involving trance as an individual activity.

In South America, the dominant form of altered state is again trance. Henney (1967:7) has noted that there is a much wider distribution and use of drugs to induce trance in South America than in North America. Trance is most frequently sought by shamans and shaman candidates, and societies of trancers are not reported. Such shamans are generally men.

As I have reported elsewhere (Bourguignon, 1968:20), we attempted to explore the relationship between drugs and altered states of consciousness for the Americas. Combining the data for North and South America, we found that there exists a statistically significant relationship (at the 0.001 level) between drugs and trance as opposed to possession trance. This is true although in most societies neither trance nor possession trance usually depend on the use of drugs. But where drugs are used, this is more likely to be the case in connection with trance. Such a linkage "makes sense" since possession trance requires greater awareness of the physical environment and of body coordination, to be able to act out the role of the spirits, as in dancing. It also requires awareness of social cues for one to act as the mouthpiece of the spirits. The linkage of drugs to hallucinatory trance as contrasted to possession trance has an almost experimental verification in a case outside American Indian societies: in Haiti, in an Afro-American setting, there is widespread possession trance, but no formalized trance and no drug use. Francis Huxley (1966) reports that he gave a Haitian vodoun priestess, much practised in possession trance, some LSD. The woman did not go into possession trance but hallucinated an appearance by one of her principal spirits. The communication made by the spirit to her was of the kind she might have received during a dream or which the spirit, in the person of the subject, might have made to others. Thus, although the cultural content determined by mythology was not modified by the drug, the form was. The altered state triggered by the drug was atypical to the customary ritual framework.

Africa

Gussler (1973) has summarized a large body of information dealing with possession trance among the South African Nguni, including among others the Zulu, the Swazi, and the Xhosa. Several different sorts of possession trance states exist among these people; most typical, however, is the cult of the diviners, who at present are, for the most part, women. Their supernatural call is generally manifested in illness which is then diagnosed as spirit possession by an established diviner. The spirit is induced to manifest himself in possession trance and to state his wishes. The patient is cured of her illness by being initiated into the cult and the possessing spirit, or rather possession trance, then appears only during ceremonies. She may become simply a cult member or may develop into a diviner and eventually head a cult group of her own. This pattern is very similar to one which is also found in a somewhat broken distribution among a series of East African peoples, and is not limited to Bantus. Beyond the area defined by Murdock (1967) as Sub-Saharan Africa, it also occurs, in the form of the zar cult, among the Moslem Somali and the Christian Amhara, as well as further to the north in the Sudan and in Egypt. I. M. Lewis (1971) has referred to this type of cult as an "amoral peripheral cult". He argues that such cults reflect the deprived situation of women, and that they involve, among other things, the supernatural coercion by the women of their menfolk. I have referred to this sort of cult as of the "East African type" (Bourguignon, 1974) although it also occurs among such West African people of the Moslem Sudan as the Hausa. Cults of this type are characterized not only by possession trance but also by the special role of possession illness as the precipitating factor in seeking membership in the cult. On the other hand, the spirit is not exorcised in this situation but an accommodation is worked out, so that the former patient becomes a devotee or even a professional diviner or cult head, active in the curing of others.

These cults may be contrasted with what I have called the "West African type" (Bourguignon, 1974) and the descendants of which we find among Afro-Americans in the Caribbean area and in Brazil (Lerch, 1972). This is in some respects closer to I. M. Lewis' "central morality cult" (Lewis, 1971). The spirits worshipped are the principal gods of the society or the ancestors of the dominant clans. (However, we may note parenthetically that in the East African type of cult the spirits which possess the Zulu diviner may well be her own ancestors.) Illness may be a cause for entering such a cult, but it is thought of as "sent" by the gods, and it is believed that initiation into the cult will cure it. In this case the spirit is not causing illness by possessing the individual but by sending it, and spirit possession trance is not induced as part of the diagnosis. The spirit is neither exorcised nor come to terms with, but installed, and the novice acquires the proper possession trance behavior. Initiation may be long, in Dahomey as long as a year (Herskovits, 1938; Verger, 1957). Both of these types of cults have groups made up of a majority of female possession trancers, and in some East African societies the membership may be entirely limited to women. Possession trance appears sometimes spontaneously, sometimes it is induced with singing and drumming and dancing, but with a few notable exceptions there is little clear evidence of the use of drugs.

These types of possession trance are not the only ones found in Africa. Among the Nuer, for example, Evans-Pritchard (1956) tells us of male prophets and among the Nuba Nadel (1946) also reports on the presence of male prophets. The diviners among the Ganda (Roscoe, 1911) were men, and the shrine priests of Ghana are apparently also men (Field, 1960). Nonetheless, we may take these female dominated cults as typical of Africa in view of their frequency and wide distribution throughout various parts of the continent.

In addition to possession trance, we also find various forms of trance in Africa. We have already mentioned the example of the Azande. Among the Fon of Dahomey and among the Yoruba (as well as many of their Afro-American descendants) non-possession trance (T) may occur as part of the initiation ritual, and there it is interpreted as a form of ritual death, the death of the old personality. (Verger, 1957, shows striking photographs of some of the rituals and of the condition of the participants.) Trance, in divination and healing, also occurs among the !Kung Bushmen (Marshall, 1962; Lee, 1968).

There is little reference to the use of drugs to induce ASCs in African religious contexts, and where there are such references to "medicines" (as in the case of Evans-Pritchard cited above) these are often so vague that they are difficult to evaluate. It may well be that drug use is relatively rare in Sub-Saharan African religious ritual, at least as compared with the Americas.

However, in recent years several studies have appeared which shed somewhat more detailed light on African drug use. One of these is the work of James Fernandez on two drug using cults among the Fang of Gabon (1972). He describes two syncretistic and relatively recent cults, Bwiti and MBiri. Bwiti is predominantly male and concerned with relations with the ancestors. MBiri is predominantly female, and primarily a curing cult. Both use several drugs, but particularly the hallucinogen *Tabernanthe iboga*. In Bwiti, heavy doses of the drug are limited to initiation into the cult, with smaller amounts taken regularly at ceremonies. Possession is not expected, that is, under the action of the drug an alteration of consciousness is expected to occur, including visions, and with lighter doses, modifications of body perceptions but not the acting out of other, spirit, personalities. When such possession trances do occur during Bwiti ceremonies, Fernandez (1972:250) says that is considered to be "nonsensical", and sometimes, undesirable and the "result of an individual's imperfections". Yet possession trance does occur, under the influence of small quantities of the drug, "particularly in women". By contrast, the female dominated MBiri curing cult where larger quantities of the drug are taken regularly, "takes just the opposite view and regards any instance of possession as having high positive value". Thus, both the male dominated and the female dominated cults use drugs, the female cult using larger doses of the hallucinogen. The cults differ not only in membership and ritual but also in aim. In the male cult the emphasis is on visions and on trips to the land of the dead, facilitated by the drug. Possession trances are undesirable but do occur, primarily among female members. The women's cult on the other hand, considers possession trance desirable, since it permits contact with the spirits causing illness.

Another example of African use of a hallucinogenic drug, in this case *Datura fastuosa* , is reported by T. G. Johnston (1973) for the Shangana-Tsonga. It is used as part of girls' initiation rites together with specific drumming patterns, energetic dancing and several other factors to produce visual and auditory hallucinations. The visions expected, and reported, deal with assurances of fertility, which are indeed the aim of the girls' initiation ceremonies. This is the only example of the use of hallucinogens in girls' initiations of which I am aware. However, although trance, as here, is reported during puberty rituals (albeit usually boys' rituals) I do not recall a case of possession trance as part of this process.

A third example comes from still another part of Africa, from the Hausa. Jacqueline Monfouga Nicolas (1972) reports that in the Bori cults, a curative possession trance cult predominantly of women, a psychotropic drug *Datura metel L.* is used. It is employed during a portion of the initiation rites in which the candidate is expected to be possessed by black (evil) spirits that are exorcized to obtain the patient's cure. The possession trances during these rituals are numerous and violent. The following days, when white (good) spirits are ritually called to possess the patient, the drug is not used. In spite of these differences, the author notes striking similarities between drug-induced and non-drug-induced possession trances, and suggests that the association of music and drug leads to a conditioned reflex, so that in the later stages of initiation (or cult learning) the music itself suffices to produce the patterns of behavior experienced first under the influence of the drug.

These three studies, which are summarized here in but the briefest terms, suggest that African drug use is, in many ways, quite different from that found in the Americas. In two cases of four it is linked to trance and in two to possession trance. In the latter cases, the subjects are women. Where possession trance occurs in the male dominated Bwiti cult of the Fang, it is considered undesirable. An association between women and possession trance is somewhat more strongly confirmed here than an association between men and hallucinogens, or between hallucinogens and trance. As far as the relationship between hallucinogens and one or the other form of altered state is concerned, another element about which we have insufficient information would appear to be significant: the strength and quantity of the drug. One would expect a strong dosage to produce visions and make the bodily coordination and mental alertness to cues required of possession trance impossible.

In Africa drugs may also be used to bring people out of an altered state rather than to induce the state. Thus, for the Ambo, Stefaniszyn (1964:160) tells of various roots (e.g., cisoko) which are soaked and drunk by a person to terminate possession trance The substance, however, is not identified adequately. Sometimes, it is the possesion spirits that take drugs or drink alcohol, i.e., the possession trance is induced prior to the taking of the drug, thus the drug cannot be said to cause the state. For example, among the Lovedu (Krige and Krige, 1943:245) a possession trancer may smoke hemp, requested by the spirit.

This addition of a drug taken in secular life to achieve an altered state when a sacred ASC is already in progress might well be worthy of physiological as well as cultural study. The preceding discussion of the relationship between trance and drugs which is so important in the Americas but relatively rare in Africa suggests that there may well exist an important nexus between these two variables and their geographic distributions. La Barre (1972:271) points out that "whereas New World natives knew eighty to a hundred such (psychotropic) drugs, the Old World had only about a half-dozen. There is good reason to expect the reverse to be true". And he goes on to explain this greater knowledge in the Americas as due to "the ubiquitous persistence of shamanism in aboriginal hunting peoples of the New World" (p. 272). And, he continues, "... it should be noted that ecstatic-visionary shamanism is, so to speak, **culturally programmed for an interest in hallucinogens and other psychotropic drugs**". (Second part of sentence was italicized in original.) He stresses, furthermore, that Old World religions, however much they were originally also based on the shamanism of hunting peoples, were massively modified by the Neolithic Revolution. And the great majority of African societies in our sample, with their possession trance patterns, are agriculturalists and/or pastoralists. And, indeed, La Barre's statement is confirmed by our findings: our statistics show an association of trance with small-scale hunting and gathering societies, of possession trance with more complex agricultural and pastoral societies. Thus, for a world-wide sample of 302 societies, we found trance much more likely to be present than possession trance among societies depending heavily on a combination of hunting, gathering and fishing for their subsistence as compared to those whose subsistence is derived from other sources (agriculture and/or animal husbandry). This association, as measured by X^2 was found to be significant below the 0.001 level of probability. The reverse of this, that is, a heavy reliance on agriculture, was found to be associated with possession trance, again at a level of probability below the 0.001 level. Note also that a heavy reliance on hunting, gathering and fishing is found among 77 per cent of the North American societies coded in Murdock's 'Ethnographic Atlas', among 28 per cent of South American societies, but only among 4 per cent of the societies of Sub-Saharan Africa (see Bourguignon & Greenbaum, 1973: Table 6).

In the preceding discussion we have stressed the differences in ASCs among geographic regions, related undoubtedly in part to ecological factors as well as to diffusion. However, as is apparent in the above, there are also important interrelationships between sacred ASCs and economic and social-structural factors. We have found that trance types (that is, societies grouped by whether they are characterized by T, T/PT, or PT) are significantly related to societal complexity, as measured by food production, agriculture, social stratification and jurisdictional hierarchy, and this is the case even when we control for diffusion (Bourguignon & Evascu, n.d.). This suggests that the differences we have described above between the patterning of ASCs in Sub-Saharan Africa and the Americas illustrate not only how societies in different regions of the world, but also societies at different levels of complexity, make use of the universal human capacity to experience ASCs. We have so far addressed ourselves to a consideration of a typology of societies as characterized by different kinds of sacred ASCs; we have also discussed some of the cultural contexts and correlates of such states. We may now turn briefly to the question of the ritualization of ASCs.

A. F. C. Wallace (1966) has defined ritual as behavior characterized by mechanical, stereotyped repetitiveness separated from "necessary" (e.g., technological) activities. The primary function of such behavior is the use of energy to communicate, yet it is "communication without information". That is, because of its stereotypical nature, ritual behavior is totally predictable, it tells us nothing new. Yet, according to Wallace, though ritual lacks informational content, it does convey two kinds of meaning: a statement of intention and a "statement of the nature of the world in which the intention is to be realized" (Wallace, 1966: 237). Specifically, ritual conveys the "image of a simple and orderly world" (p. 239). Ritual is religious ritual only when a supernatural premise or myth is used to rationalize it. Other, secular, rationalizations, or myths are also possible, and at least in the case of animals, so is ritual behavior without (verbal) rationalization.

Wallace, furthermore, speaks of what he terms the "ritual reorganization of experience" which he considers to be a kind of learning and which reduces the complex world of experience to an orderly world of symbols. This learning process, he tells us, involves five stages, rather similar to the stages of the rites of passage defined by van Gennep. These stages are: pre-learning, separation (of the ritualist from "irrelevant environmental information"), suggestion (by which new cognitive structures are acquired), execution (that is, the acting in accordance with these new cognitive structures) and finally, maintenance (of the new, transformed identity).

Wallace applies this sequence, in the examples he gives, to initiation rituals (rites of passage), drug trances, "possession states", and conversion experiences. However, he appears to wish to set out a much broader pattern, valid also for rituals which do not involve ASCs. And he explicitly notes that he considers this sequence valid for secular as well as religious rituals.

To what extent can we generalize about ASCs in this manner? Wallace says all rituals, as defined by him, involve these stages. Do all ASCs involve ritual in this sense, or more narrowly, do all religiously institutionalized ASCs, coded by us as T, or PT, or some sequence thereof, fit this pattern?

We may begin by considering the Haitian case as an example on which the model may be tested. The Haitian folk religion of 'vodoun' is well known for the importance that possession trance plays both within its ceremonies and in its belief system. It is believed that the world is peopled by spirits, termed various 'saints', 'loa', or 'mysté' and that these may temporarily 'mount' their faithful, the personality of the spirit displacing that of its human host, or 'horse' in local parlance. During ceremonies, the spirits are called by means of drum rhythms, songs, dances, ritual paraphernalia, etc. and given persons may respond to these cues by going into an altered state and acting out the appropriate spirit role, that is, they go into a possession trance. Afterwards, they are somewhat disoriented and are expected to be (and often are) amnesic for the events of the 'possession'. In terms of Wallace's pattern, there is pre-learning: the subject has experienced the state before and knows what to expect, and there is broad general knowledge concerning spirits, their identities and their behaviors. There is separation: the drums, dances etc. shut out mundane matters and help the individual concentrate on what is expected of him (or her). This is followed by the state itself, which is referred to by Wallace as suggestion: the psychological materials relevant for the role of the spirit is mustered and acted out. Here 'execution' also occurs during the altered state, and it is not clear to me how the two are distinguished. The 'spirit', however, may express wishes and give orders, both to the 'horse' and to others and in this sense 'execution' of ritual work will have an impact over a longer period of time. Maintenance: Wallace, himself, suggests that in ritual possession there is no maintenance beyond the trance state, except to say that apparently persons having experienced such a state have a lower threshhold for dissociation. It does indeed appear that persons will go into the state more easily the greater their habituation to the stimuli which induce it. Yet, at the same time, we also know that as people get to be older, and also, the Haitians say, as they acquire greater esoteric knowledge and more control over the spirits, they will be possessed less often. An interesting point should be noted here with regard to the notion of 'stereotype' of ritual behavior: The induction of the ASC ('separation') is indeed stereotyped: spirits are called with their own particular rhythms, motions, attributes, etc. However, the behavior of the possession trancer is likely to be stereotyped only to a limited degree. That is, each spirit has his own known

characteristics, yet the same spirit 'in the head of' two different people will act slightly differently. Also, on any given occasion, the behavior of the spirit cannot be fully anticipated. That is, the spirits are called precisely in order to establish informational two-way communication with them. They are asked questions, asked for advice, requested to help; they give orders, express opinions, etc. Are we to say, then, that this is no longer 'ritual' behavior because stereotype does not pertain to this phase of the larger ceremony? It should be noted here that societies that have possession trance vary widely precisely in regard to this point. In Haiti there is a great deal more innovation and unpredictability apparently than in the parallel and ancestral cults of the Yoruba and the Fon. Yet there, too, there is a good deal of improvisation, as shown for example in the writings of P. Verger (1969). R Horton (1969) differentiates several types of spirit possession among the Kalabari of West African, in some of which the behavior is highly stereotyped, or one might say 'ritualized', while in others the actors have considerable latitude in their behavior. The implications of these differences in the degree to which the behavior of actors in possession trance states is stereotyped or the degree to which they are free to innovate or improvise is an important variable, I believe, which it would be worthwhile to investigate more closely on a comparative, cross-cultural basis. I would suggest that those limited to stereotyped behavior are likely to have a conservative effect in the maintenance of ritual and belief and of their social context. On the other hand, where the actor can improvise and speak with authority, he may be able to become a force for innovation. The same is probably true with regard to visionary states, where we would wish to distinguish between the types of visions experienced, the kind of dream-learning undergone, the kinds of messages and instructions received, etc. Often enough, a first, spontaneous experience, based, it is true, on pre-learning, but not ritually induced, may lead to the development of a ritual pattern, imposing on others a ritual, stereotyped reenactment of what was a genuine and spontaneous altered state in the case of a prophet or founder. In the process, ASCs may come to be mimed rather than to be experienced.

In the context of a discussion of the ritual character of ASCs, it is interesting to note that at present, in many parts of the world, what was once religious ritual incorporating such states is now often becoming a theatrical reenactment of such states in what is essentially secular ritual. Haitian 'vodoun' ceremonies are stages for tourists as are Balinese temple dramas and kris dances. This, of course, is nothing new, for Jane Harrison long ago pointed to the roots of Greek drama in Dionysian ritual. To the extent that the theatrical performance has a completely prescribed set of actions and texts it may, indeed, be more stereotyped, thus more highly ritualistic, than a religious ceremony (with or without ASC) which may have an unpredictable element. In this sense, the little interpersonal (and even, inter-spirit) scenes or playlets acted out in Haitian 'vodoun' ceremonies are more commedia dell' arte than drama!

So far, we have been dealing with intentionally and ritually-induced ASCs. Where such states occur spontaneously, Wallace's stages of ritual learning may still be applicable to a limited extent: the spontaneous behavior and/or experience will incorporate cultural as well as idiosyncratic cognitive and affective aspects of the individual and thus it is appropriate to speak of pre-learning here as well. The precipitating factors, whether drugs, sudden shock or crisis, or whatever, may be expected to fit the requirements of the state of 'separation', although here they are not planned by cult leaders or incorporated (as yet) into traditional ritual patterns. The experience may involve a 'resynthesis', in Wallace's terms, in a stage of suggestion and may have effects in stages of execution and maintenance. In the case of the Seneca prophet, Handsome Lake, as Wallace has told us, and in the case of many other prophets, the foundation may be laid for a new social movement in such experiences. However, where highly idiosyncratic, uncontrolled behavior results, social efforts are likely to be undertaken to bring this behavior under control. Such control is established by the development of a myth to explain the behavior and the creation of ritual means of coping with it. That is, (largely) stereotyped actions under controlled conditions usually heavily laden with symbolism, and thus with unconscious as well as conscious messages, are called upon to deal with such situations. The action may involve exorcism of evil spirits posited to account for the behavior as in the Judeo-Christian tradition; initiation as in the East African cults we discussed earlier may be required. The killing of an amok individual may be thought of in these terms. In modern societies, often enough, methods of dealing with spontaneous ASC may involve defining a hallucinating individual as sick and applying the consequences of that definition: hospitalization, psychiatric treatment, etc. And this, too, may involve a range of stereotyped and predictable, thus ritual, actions, albeit ritual action rationalized by a secular ideology.

I would like to conclude by repeating my earlier observation that ASCs occur in all human societies. They are very frequently embedded into religious patterns of belief and ceremonial, with varying degrees of ritualization. They are also very often ritualized in a secular context. Ritualization may be thought of as an imposition of order, a bringing under social and ideological control of what are potentially disruptive psychological states and forces. The very frequency of ritualization of ASCs suggests that in most societies many kinds of such states are so viewed, in a traditional sacred context, as well as in a modern secular one.

BIBLIOGRAPHY

Barry, Herbert III, I. L. Child & M. K. Bacon 1959, Relation of child training to subsistence economy. American Anthropologist, 61:51-63.

Bourguignon, Erika, 1968, A cross-cultural study of dissociational states; final report. Columbus, Ohio: The Ohio State University Research Foundation.

Bourguignon, Erika, 1972, Dreams and altered states of consciousness in anthropological research. In F. L.K. Hsu (ed.), Psychological anthropology (new edition). Cambridge, Mass.: Schenkman.

Bourguignon, Erika, 1973, Introduction: a framework for the comparative study of altered states of consciousness. In E. Bourguignon (ed.), Religion, altered states of consciousness and social change. Columbus, Ohio: Ohio State University Press.

Bourguignon, Erika, 1974, Illness and possession: elements for a comparative study. R. M. Bucke Memorial Society Newsletter-Review, 7.

Bourguignon, Erika & T. L. Evascu, n.d., Altered states of consciousness within a general evolutionary perspective: a holocultural analysis. Behavior Science Research (forthcoming).

Bourguignon, Erika & L. S. Greenbaum, 1973, Diversity and homogeneity in world societies. New Haven: HRAF Press.

Evans-Pritchard, E. E., 1937, Witchcraft, oracles and magic among the Azande. Oxford: Clarendon Press.

Evans-Pritchard, E. E., 1956, Nuer Religion. Oxford: Clarendon Press.

Evans-Pritchard, E. E., 1962 (orig. 1936), Zande theology, social anthropology and other essays. New York: Free Press of Glencoe.

Fernandez, J. W., 1972, Tabernanthe Iboga: Narcotic ecstasis and the work of the ancestors. In P. T. Furst (ed.), Flesh of the Gods. New York: Praeger.

Field, M. J., 1960, Search for security. Evanston, Ill.: Northwestern University Press.

Fischer, Roland, 1971, Cartography of the ecstatic and meditative states. Science, 174:897-904.

Fischer, Roland, 1974, Cartography of inner space. In R. K. Siegel & L. J. West (eds.), Hallucinations: behavior, experience, theory. New York: Wiley.

Gussler, Judith, 1973, Social change, ecology and spirit possession among the South African Nguni. In E. Bourguignon (ed.), Religion, altered states of consciousness and social change. Columbus, Ohio: Ohio State University Press.

Hallowell, A. I., 1955, Culture and experience. Philadelphia: University of Pennsylvania Press.

Hallowell, A. I., 1958, Ojibwa metaphysics of being and the perception of persons. In R. Tagiuri and L. Petrullo (eds.), Person perception and interpersonal behavior. Stanford: Stanford University Press.

Henney, J. H., 1967, Cultural context of possession and trance states in South America. Working paper 14, Cross-cultural study of dissociational states. Department of Anthropology, The Ohio State University.

Herskovits, M. J., 1938, Dahomey: an ancient West African kingdom. New York: J. J. Augustin.

Horton, Robin, 1969, Types of spirit possession in Kalahari religion. In J. Beattie and J. Middleton (eds.), Spirit mediumship and society in Africa. New York: Africana Publishing Corp.

Huxley, Francis, 1966, The invisibles: Voodoo Gods in Haiti. New York: McGraw-Hill.
Johnson, T. G., 1973, Believed fertility through altered states of consciousness among the Shangana-Tsonga of Southern Africa. Transcultural Psychiatric Research Review, 10:152-59.
Krige, J. D. & E. J. Krige, 1943, The realm of a Rain-Queen. London: Oxford University Press.
La Barre, Weston, 1972, Hallucinogens and the Shamanic origins of religion. In P. T. Furst (ed.), Flesh of the Gods. New York: Praeger.
Lee, R. B., 1968, The sociology of !Kung Bushman trance performances. In R. Prince (ed.), Trance and possession states. Montreal: R. M. Bucke Memorial Society.
Lerch, Patricia, 1972, The role of women in possession trance cults of Brazil. Unpublished M.A. thesis, Department of Anthropology, The Ohio State University.
Lewis, I. M., 1971, Ecstatic religion. Baltimore: Penguin Books.
Ludwig, Arnold M., 1968, Altered states of consciousness. In R. Prince (ed.), Trance and possession states. Montreal: R. M. Bucke Memorial Society.
Ludwig, Arnold M., 1972, (orig. 1966), Altered states of consciousness. In C. T. Tart (ed.), Altered states of consciousness. Garden City, New York: Anchor Books, Doubleday.
Marshall, Lorna, 1962, !Kung Bushman religious beliefs. Africa, 32:221-252.
Monfouga-Nicolas, Jacqueline, 1972, Ambivalence et culte de possession. Paris: Editions anthropos.
Murdock, G. P., 1967, Ethnographic atlas. Pittsburgh: University of Pittsburgh Press.
Nadel, S. F., 1946, A study of Shamanism in the Nuba Mountains. Journal of the Royal Anthropological Institute, 76:25-37.
Price-Williams, D. R. (ed.), 1970, Cross-cultural studies: Selected readings. Baltimore: Penguin Books.
Roscoe, John, 1911, The Baganda. London: McMillan.
Segall, M. H., D. Campbell & M. J. Herskovits, 1966, The influence of culture on visual perception. Indianapolis: Bobbs-Merrill.
Stefaniszyn, B., 1964, Social and ritual life of the Ambo of Northern Rhodesia. London: Oxford University Press.
Swanson, G. E., 1973, The search for a guardian spirit: a process of empowerment in simpler societies. Ethnology, 12:359-378.
Verger, Pierre, 1957, Notes sur le culte des Orişa et Vodun. Mémoires d'IFAN, Dakar.
Verger, Pierre, 1969, Trance and convention in Nago-Yoruba spirit mediumship. In J. Beattie and J. Middleton (eds.), Spirit mediumship and society in Africa. New York: Africana Publishing Corp.
Wallace, A. F. C., 1966, Religion: an anthropological view. New York: Random House.

RICHARD EVANS SCHULTES 2

THE BOTANICAL AND CHEMICAL DISTRIBUTION OF HALLUCINOGENS*

Out of the vast array of species in the plant kingdom — variously estimated at from 200,000 to 800,000 — a few have been employed in primitive societies for millennia to induce visual, auditory, tactile, and other hallucinations Because of their unearthly effects that often defy description, they have usually been considered sacred and have played central roles as sacraments in aboriginal religions (Schultes, 1969c).

Scientific interest in hallucinogenic agents has recently been intense, partly because of the hope of finding potentially valuable drugs for use in experimental or even therapeutic psychiatry and also for employment as possible tools in an explanation of the biochemical origins of mental abnormalities (Hoffer & Osmond, 1967).

While psychoactive species are widely scattered throughout the plant world, they appear to be concentrated more or less amongst the fungi and the angiosperms. The bacteria, algae, lichens, bryophytes, ferns, and gymnosperms seem to be notably poor or lacking in species with hallucinogenic properties (Schultes, 1969-70). These hallucinogenic properties can be ascribed, likewise, to only a few kinds of organic constituents, which may be conveniently divided into two broad groups: nitrogenous and non-nitrogenous compounds (Der Manderosian, 1967a; Farnsworth, 1968, 1969; A. Hofmann, 1961a, 1968; W. I. Taylor, 1966; Usdin & Efron, 1967). See Figure 1 for basic chemical skeletons.

The nitrogenous compounds play by far the greater role and comprise, for the most part, alkaloids or related substances, the majority of which are, or may be biogenetically derived from, the indolic amino acid tryptophan. They may be classified into the following groups: 1. β-carbolines; 2. ergolines; 3. indoles; 4. isoquinolines; 5. isoxazoles; 6. β-phenylethylamines; 7. quinolizidines; 8. tropanes; 9. tryptamines. Non-nitrogenous compounds, which are the active principles in at least two well known hallucinogens include: 1. dibenzopyrans; and

* Reprinted, with permission, from 'Annual Review of Plant Physiology', Vol. 21.

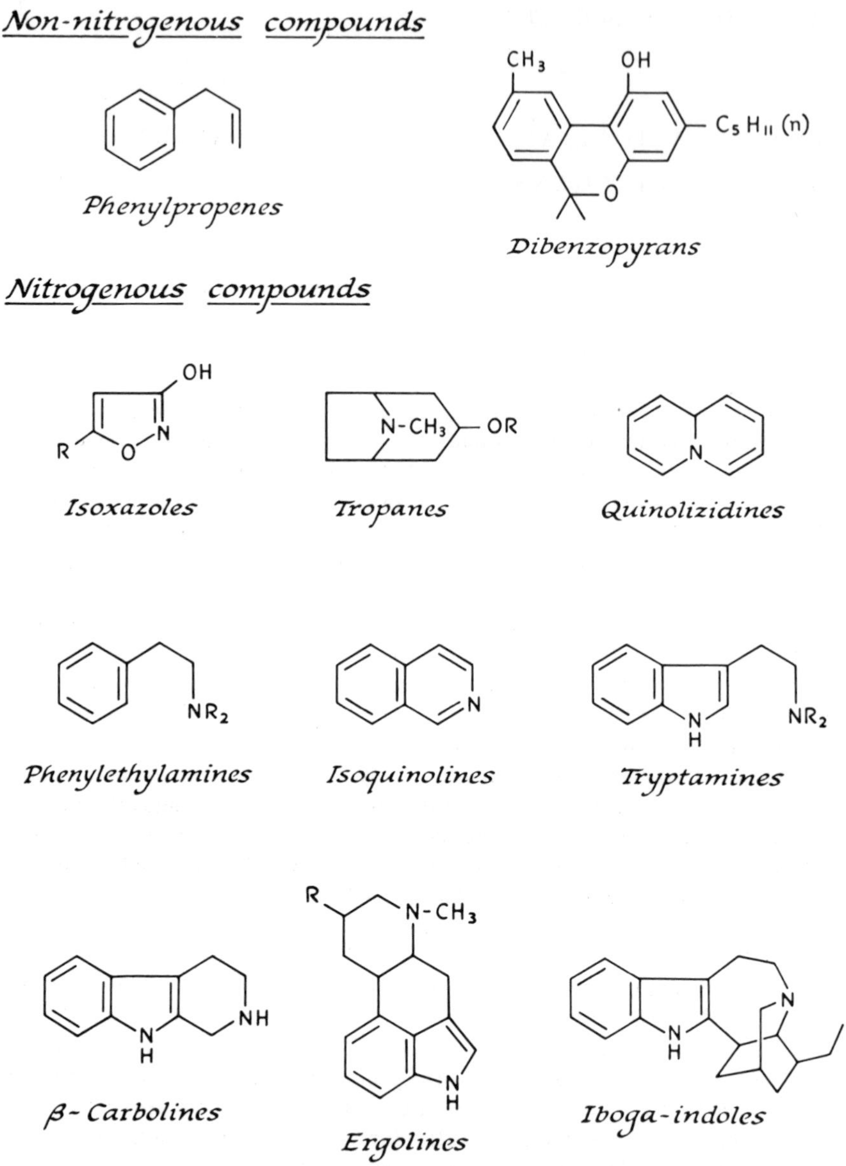
Non-nitrogenous compounds
CH3
OH
C5 H11 (n)
O
Phenylpropenes
Dibenzopyrans
Nitrogenous compounds
OH
N
R
O
N-CH3
OR
N
Isoxazoles
Tropanes
Quinolizidines
NR2
N
N
H
NR2
Phenylethylamines
Isoquinolines
Tryptamines
R
N-CH3
N
H
NH
N
H
N
N
H
β- Carbolines
Ergolines
Iboga-indoles

Figure 1. Basic chemical skeletons of principal hallucinogens

2. phenylpropenes; other compounds, such as catechols and alcohols, may occasionally play a role.

In the study of hallucinogenic plants, two considerations must be borne in mind. One consideration reminds us that there are some of these psychoactive plants used in primitive societies for which the active chemical principles are as yet not known. The other emphasizes that man undoubtedly has utilized only a few of the species that actually do possess hallucinogenic principles: we are, as yet, far from knowing how many plants are endowed with psychotomimetic constituents, but there are certainly many more than the few employed by man as hallucinogens (Schultes, 1967a).

While almost all hallucinogenic compounds are of vegetal origin, a few may be wholly or partly synthetic. The potent hallucinogen, lysergic acid diethylamide (LSD), although very closely allied chemically to the naturally occurring ergolines, has not been found in the plant kingdom.

NON-NITROGENOUS PRINCIPLES

1. Dibenzopyrans

Cannabaceae — *Cannabis.* The most important of the non-nitrogenous hallucinogens are the dibenzopyrans found in *Cannabis sativa,* source of marijuana, bashish, bhang, ganja, and other narcotic products. *Cannabis,* a monotypic genus sometimes placed in the Moraceae but often allocated, together with the hops plant (*Humulus*), in a separate family, Cannabaceae, represents perhaps one of the oldest and certainly the world's most widespread hallucinogen. The source also of hempen fibres and of an edible seed-oil, the plant is native probably to central Asia but is now found cultivated or spontaneous in most parts of the world. It is represented by many agricultural varieties and ecological races or strains, some of which are rich, some poor, some even lacking in the intoxicating principles.

The biodynamic activity of *Cannabis* is due to a number of constituents contained in a red oil distilled from the resin, mainly to a mixture of stereoisomers collectively called tetrahydrocannabinols and sundry related compounds, including cannabinol, cannabidiol, cannabidiolic acid, cannabigerol, and tetrahydrocannabinolcarboxylic acid (A. Hofmann, 1968; Korte & Sieper, 1965; Schulz, 1964). The compound Δ^1-tetrahydrocannabinol has recently been shown to be the principal biologically active constituent of *Cannabis* (Mechoulam, Braun & Gaoni, 1967; Mechoulam & Gaoni, 1965). Cannabichrome has likewise been reported as active. Cannabinol and cannabidiol are devoid of euphoric properties, although cannabidiol, when heated with an acidic catalyst, may be converted into an active mixture of tetrahydrocannabinols, a conversion that may be effected during the smoking of *Cannabis.* Cannabigerol and cannabidiolic acid are sedative, and the latter compound

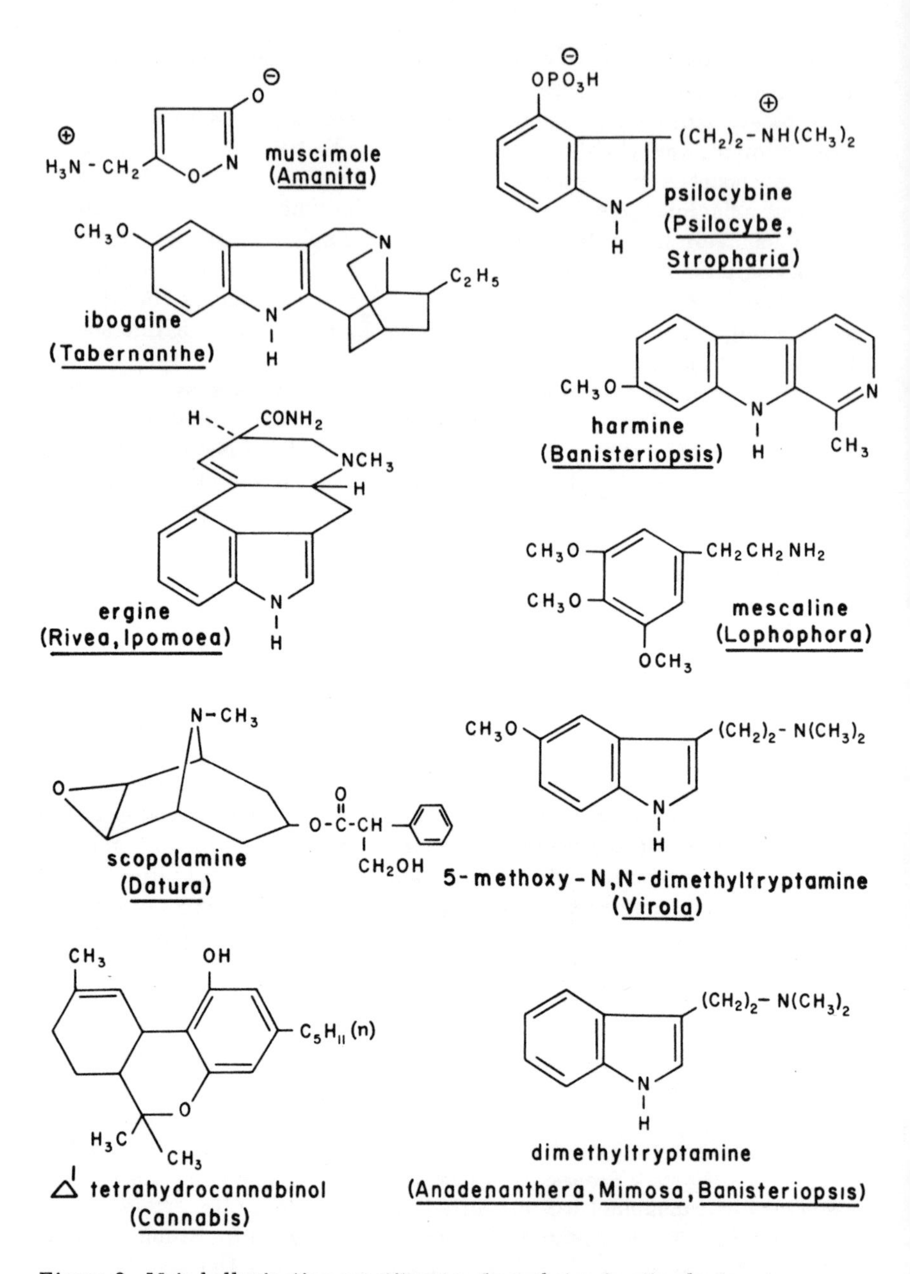

Figure 2. Main hallucinating constituents of psychotomimetic plants

has antimicrobial properties (Farnsworth, 1968; A Hofmann, 1968). Cannabinol and the tetrahydrocannabinols have been synthesized. The first biologically active principle to have been structurally elucidated and synthesized is Δ^1-2, 3-trans tetrahydrocannabinol (A. Hofmann, 1968).

The widely recognized and extreme variation in psychoactive effects of *Cannabis* is due possibly to the instability of some of the constituents which, upon aging or maturation of the plant, may be converted from active to inactive compounds or vice versa. It is believed that some of these conversions take place more readily and rapidly in the drier tropics than in temperate climates. Many phytobiotic factors seem to have effects on the chemical composition of the resin.

This variability and the absence of controlled experiments — almost all experimentation has been done with crude plant material of unknown chemical constitution — have led to a disturbing lack of uniformity of opinion on the physiological effects of *Cannabis,* a situation that has seriously handicapped social and legal control of the use or abuse of the drug (Weil, 1969b).

2. Phenylpropenes

Myristicaceae — *Myristica.* The tree that yields the spices nutmeg and mace — *Myristica fragrans* — is thought to have been employed aboriginally as a narcotic in southeastern Asia, where it is native. It is sometimes used as an hallucinogen in sophisticated circles in Europe and North America and has occasionally become a problem in prisons in the United States (Weil, 1965, 1967, 1969a).

Although its toxicology has not yet been wholly elucidated, the psychoactive principles are contained probably in the essential oil of the seed and aril. The composition of nutmeg oil is highly variable, both qualitatively and quantitatively, but it does contain fatty acids, terpenes, and aromatics The psychopharmacological effects may be attributable to several phenylpropenes. Elemicine, myristicine, and safrol have been suggested as the active constituents of the oil, which may also contain eugenol, isoeugenol, methylisoeugenol, methoxyeugenol, methyleugenol, and isoelemicine. It seems doubtful that myristicine or safrol are responsible for a significant part of the hallucinogenic effects. While these properties may be attributable largely to elemicine, no studies on the psychopharmacological activity of pure elemicine or safrol have as yet been made (Shulgin et al., 1967; Truitt, 1967).

For hallucinating purposes, ground nutmeg is taken orally in large doses, usually several teaspoonfuls. The effects vary appreciably but are often characterized by distortion of perception of time and space, dizziness, tachycardia, dry mouth, headache, and occasionally visual hallucinations (Weil, 1967).

Myristica is a genus of some 120 species of the Old World tropics.

The only commercially important species is *M. fragrans,* native of the Moluccas and source of two products: nutmeg from its seed and mace from the aril surrounding the seed.

3. Other compounds: alcohols

Labiatae — *Lagochilus.* For centuries the Tajik, Tartar, Turkomen, and Uzbek tribes of central Asia have used *Lagochilus inebrians* as an intoxicant. The leaves, gathered usually in October, are toasted and made into a tea, sometimes with stems, flowering tops, and the white flowers. Honey or sugar is added to lessen the bitterness (Bunge, 1847).

In 1945, a crystalline material called lagochiline, at first thought to be an alkaloid, was isolated. More recent studies, however, have indicated that it is a polyhydric alcohol and that it occurs in concentrations of up to 3 per cent of dried plant material (Tyler, 1966).

Because of its versatile effects, *Lagochilus inebrians* was made official in the eighth edition of the 'Russian Pharmacopoeia'. The recognized sedative activity of the plant is due possibly to the same constituent responsible for the central nervous system activity basic to its folk use as a narcotic (Tyler, 1966). The genus *Lagochilus* comprises some 35 species occurring from central Asia to Persia and Afghanistan.

NITROGENOUS PRINCIPLES

1. β-Carbolines

Zygophyllaceae — *Peganum.* The Syrian rue or *Peganum harmala* is an herb found in dry localities from the Mediterranean area east to India, Mongolia, and Manchuria. It is a member of a genus of six species distributed in dry areas of Asia Minor and Asia and in southwestern United States and Mexico. Although this and other species of *Peganum* have long been esteemed in folk medicine, its purposeful employment as an hallucinogen is open to question, vague reports notwithstanding, even though it does have psychotomimetic principles (Porter, 1962).

The seeds of *Peganum harmala* contains harmine, harmaline, harmalol, and harman, bases of a typical β-carboline structure of wide botanical and geographical distribution, having been isolated from at least eight plant families of both the New and the Old World (Deulofeu, 1967; Willaman & Schubert, 1961).

Malpighiaceae — *Banisteriopsis.* In wet tropical forest areas of northern South America, the aborigines use as hallucinogens several species of *Banisteriopsis* containing harmala alkaloids: *B. caapi, B. inebrians.* An intoxicating drink is prepared from the bark of the stems in the Amazon of Brazil, Bolivia, Colombia, Ecuador, and Peru, the Orinoco of

Venezuela, and the Pacific coast of Colombia. It is known variously as 'ayahuasca', 'caapi', 'yajé', 'natema', 'pinde', or 'dapa'. Usually only one species enters the preparation, but occasionally admixtures may be employed (Friedberg, 1965; Schultes, 1957, 1961).

A genus of some 100 species of tropical America, *Banisteriopsis* is taxonomically still rather poorly understood. This is true especially of *B. caapi* and *B. inebrians,* partly because of the lack of fertile material for study of these infrequently flowering jungle lianas, even though the first botanical attention to this drug plant dates from 1854, when it was first encountered in northwestern Brazil by the explorer Spruce (Schultes, 1968a, b).

The chemistry of these hallucinogenic species of *Banisteriopsis* has been more critically investigated than the taxonomy, yet failure of chemists to insist upon botanically determined material for analyses has created chaos. Earlier workers isolated alkaloidal constituents from plants probably referable to *B. caapi* which they named telepathine, yageine, banisterine, all of which were eventually identified as harmine (Henry, 1949). More recent examination of botanically authenticated material of this species has established the presence in the bark — and sometimes in the leaves — of harmine as well as occasional lesser amounts of harmaline and β-tetrahydroharmine (Chen & Chen, 1939; Deulofeu, 1967). Recent investigations of *B. inebrians* have isolated harmine from the stems and minute amounts of what appears to be harmaline (O'Connell & Lynn, 1953). An interesting chemical study of stems of the type collection of *B. caapi* has indicated, in spite of passage of some 115 years, the presence of harmine in concentrations matching that of freshly collected material (Schultes, Holmstedt & Lindgren, 1969).

While *Banisteriopsis caapi* is normally employed as a drink, recent indirect evidence from the northwest Amazon indicates that it may also be used as a snuff. Harmala alkaloids have been reported from snuff powders prepared from a vine said also to be the source of an intoxicating drink, but voucher botanical specimens are lacking (Holmstedt & Lindgren, 1967).

Harmine has been isolated from *Cabi paraensis* of the eastern Amazon, a genus closely allied to *Banisteriopsis.* While it is valued in folk medicine, it is apparently not employed as an hallucinogen (Mors & Zaltzman, 1954).

2. Ergolines

Convolvulaceae — *Ipomoea, Rivea.* The early Spanish chroniclers of Mexico reported that the Indians employed in their religious and magic rites an halluicogenic seed called 'ololinqui' by the Aztecs. It was also used medicinally, and when applied as a poultice was said to have analgesic properties.

Known as 'coatl-xoxouhiqui' (snake plant), it was adequately illustrated as a morning glory. Although several Mexican botanists accepted this identification during the last century, not until 30 years ago was a voucher specimen of a convolvulaceous plant, the seeds of which were employed as a divinatory hallucinogen, collected amongst the Mazatecs of Oaxaca and determined as *Rivea corymbosa.* Later field work uncovered similar uses of another morning glory, *Ipomoea violacea,* amongst the Zapotecs, also of Oaxaca; this species represents possibly the narcotic 'tlitliltzin' of the ancient Aztecs (MacDougall, 1960; Schultes, 1941; R. G. Wasson, 1962b).

In the interval, 'ololiuqui' had been identified as a species of *Datura,* an identification that gained wide acceptance (Safford, 1916b). The reasoning upon which this theory was based held that in four centuries no narcotic used had been observed for a morning glory; the convolvulaceous flowers resembled those of a *Datura* and might have led to confusion; descriptions of ololiuqui-intoxication coincided closely with that induced by *Datura; Datura* had been and still is employed as a divinatory narcotic in Mexico; and, most significantly, no psychoative principle was known from the Convolvulaceae.

Experimental psychiatry indicated that *Rivea* was definitely hallucinogenic, supporting ethnobotanical field work (Osmond, 1955). Yet chemists were unable to isolate any inebriating constituents, until 1960 and subsequently, when ergot alkaloids related to the synthetic hallucinogenic compound LSD were found in the seeds of both *R. corymbosa* and *I. violacea.*

The main psychotomimetic constituent of the seeds of both species are ergine (d-lysergic acid diethylamide) and isoergine (d-isolysergic acid diethylamide) which occur together with minor alkaloids: chanoclavine, elymoclavine, and lysergol. Ergometrine appears to be present in seeds of *I. violacea* but absent in *R. corymbosa.* The total alkaloid content of *R. corymbosa* seed is 0.012 per cent; of *I. violacea,* 0.06 per cent — and, indeed, Indians use smaller quantities of the latter than of the former (A. Hofmann, 1961a,b, 1963a,b, 1964, 1966, 1967, 1968; A. Hofmann & Cerletti, 1961).

The discovery of ergot alkaloids — constituents of *Claviceps purpurea,* a relatively primitive fungus — in one of the phylogenetically most advanced angiosperm families was unexpected and is of great chemotaxonomic interest. Suspicion that fungal spores might have contaminated the convolvulaceous seeds was ruled out experimentally (Taber & Heacock, 1962); and the discovery of these alcaloids in fresh leaves, stalks, and roots of *I. violacea* and, to a minor extent, in leaves of *R. corymbosa* indicated that these constituents are produced by the tissue of the morning glories themselves, not by infecting fungi (Taber, Heacock & Mahon, 1963).

Large amounts of a new glycoside, turbicoryn, were likewise isolated from seeds of *R. corymbosa,* but this compound apparently has no part in the psychotomimetic action (Cook & Keeland, 1962; Pérezamador & Herrán, 1960).

Studies have shown the presence of these ergot alkaloids in a number of horticultural 'varieties' of *I. violacea* and other species of *Ipomoea*, as well as in the related genera *Argyreia* and *Stictocardia* (Der Marderosian, 1967b; Der Marderosian & Youngken, 1966; Genest et al., 1965; Hylin & Watson, 1965; Taber, Vining & Heacock, 1963).

There are folklore references to psychotomimetic uses of *Ipomoea* in Ecuador, where its common names, 'borrachera' and 'matacabra', refer to its inebriating or toxic effects. Ergot alkaloids have been isolated from this species (Naranjo, 1969).

The nomenclature and taxonomy of the Convolvulaceae are in a state of extreme confusion, especially as to delimitation of genera (Der Marderosian, 1965; Schultes, 1964, Shinners, 1965). *Revea,* primarily an Asiatic species of woody vines, has five Old World species and one, *R. corymbosa,* in the western hemisphere, occurring in southernmost United States, Mexico, and Central America, some of the Caribbean islands, and the northern coast of South America. *R. corymbosa* has at least nine synonyms, of which *Ipomoea sidaefolia* and *Turbina corymbosa* are most frequently employed. *Ipomoea,* comprising upwards of 500 species in the warm temperate and tropical parts of the hemisphere, is a genus of climbing herbs or shrubs, rarely semi-aquatic. *I. violacea*, often referred to by its synonyms *I. rubro-caerulea* and *I. tricolor,* is represented in horticulture by a number of 'varieties', such as Heavenly Blue, Pearly Gates, Flying Saucers, Wedding Bells, Summer Skies, and Blue Stars — all of which contain the hallucinogenic ergot alkaloids (Der Marderosian, 1967b).

3. Iboga-indoles

Apocynaceae — *Tabernanthe.* Probably the only member of this alkaloid-rich family known definitely to be utilized as an hallucinogen is 'iboga', the yellowish root of *Tabernanthe Iboga.* This narcotic is of great social importance, especially in Gabon and nearby portions of the Congo in Africa. The use of iboga, first reported by French and Belgian explorers in the middle of the last century, appears to be spreading. In Gabon, it is employed in initiation rites of secret societies, the most famous of which is the Bwiti cult. Sorcerers take the drug before communicating with the spirit world or seeking advice from ancestors (Pope, 1969).

Twelve closely related indole alkaloids have been reported from iboga; they comprise up to 6 per cent of the dried material. Ibogaine, apparently the principal psychoactive alkaloid, acts as a cholinesterase inhibitor, a strong central stimulant, and as an hallucinogen (Hoffer & Osmond, 1967; Pope, 1969). *Tabernanthe* is a genus of about seven species native to tropical Africa.

Sometimes other plants — occasionally as many as 10 — are taken with iboga, but few have been chemically investigated. One of the most interesting, the euphorbiaceous *Alchornea floribunda,* is employed also in the same way as iboga in another secret society in Gabon, but is apparently not hallucinogenic. Its active principle seems to be the indole yohimbine (Tyler, 1966).

4. Isoquinolines

Cactaceae — *Lophophora. Lophophora williamsii,* the 'peyote' cactus, has more than 30 bases belonging to the phenylethylamines and the simple isoquinolines (Agurell, 1969; Reti, 1950, 1954). The visual hallucinations are due to the phenylethylamine mescaline (see Section 6), but other aspects of the complex peyote-intoxication, such as auditory, tactile, and taste hallucinations and other effects, may be due in part at least to the isoquinolines, either alone or in combination. Among the important isoquinolines present are anhalamine, anhalidine, anhalinine, anhalonidine, pellotine, lophophorine, peyoglutam, mescalotam, and several as yet unnamed bases recently isolated (Agurell, 1969; Der Marderosian, 1966; Kapadia & Fales, 1968a, b; Kapadia & Highet, 1968; Kapadia, Shah & Zalucky, 1968; Lundstrom & Agurell, 1967; McLaughlin & Paul, 1965, 1966; Pallares, 1960; Reti, 1950, 1954).

5. Isoxazoles

Agaricaceae — *Amanita.* While *Amanita muscaria* — the fly agaric, a mushroom of the north-temperate zone of Eurasia and North America — may represent one of the oldest of the hallucinogens used by man, only very recently has a clarification of the chemistry of its active principles begun to take shape (Heim, 1963b).

The Aryan invaders of India 3, 500 years ago worshipped a plant, the god-narcotic 'soma', center of an elaborate cult in which the inebriating juice was ceremonially drunk (R. G. Wasson, 1969). More than 1, 000 hymns to soma have survived in the Rig Veda, describing the plant and its significance in detail. The use of soma died out 2, 000 years ago. Botanists have proposed more than 100 species in attempts to identify soma, but none have been satisfactory. The most recent identification of soma as *Amanita muscaria* appears, from the indirect evidence at hand, to be highly probable.

In the 18th century, Europeans discovered the narcotic use of *Amanita muscaria* among primitive tribesmen of Siberia. Until very recently it was employed as an orgiastic or shamanistic inebriant by the Ostyak and Vogul, Finno-Ugrian peoples of western Siberia, and the Chukchi, Koryak and Kamchadal of northeastern Siberia. Tradition has established its use among other peoples (Brekhman & Sam, 1967; Lewin, 1964; Wasson, 1967, 1969).

In Siberia, several mushrooms sufficed to induce intoxication —

taken as extracts in water or milk, alone or with the juice of *Vaccinium uliginosum* or *Epilobium angustifolium.* A dried mushroom may be held moistened in the mouth or women may chew them and roll them into pellets for the men to ingest. Since the mushrooms often are expensive, the Siberians practiced ritualistic drinking of the urine of an intoxicated person, having discovered that the inebriating principles were excreted unaltered by the kidneys. Urine-drinking is mentioned also in the Rig Veda hymns to soma(Wasson, 1969).

Since the discovery in 1869 of muscarine, the intoxicating activity of *Amanita muscaria* has been attributed to this alkaloid. Recent studies, however, have indicated that muscarine represents a minor constituent of the mushroom to which the strong inebriation could hardly be attributed. Trace amounts of bufotenine in the carpophores, likewise, could not be responsible, if indeed it be present. The reported presence of tropane alkaloids has been shown to be due to incorrect interpretation of chromatographic data. Other compounds detected in *A. muscaria* are choline, acetylcholine, and muscaridine (Waser, 1967).

Recent chemical and pharmacological studies have shown that the principal biologically active constituents appear to be muscimole — the enolbetaine of 5-aminoethyl-3-hydroxy-isoxazole — an unsaturated cyclic hydroxamic acid which is excreted in the urine; and ibotenic acid — the zwitterion of α-amino-α-(3-hydroxy-isoxazoyly-(5))-acetic acid monohydrate. The less active musoazone, likewise an amino acid, α-amino-α(2(3H)-oxazolonyl-(5))-acetic acid, is present in varying but lesser amounts. Structurally related to these isoxazoles is the antibiotic oxamycine which often has psychactive effects — mental confusion, psychotic depression, abnormal behavior — in man. Other active substances structurally still not elucidated are also known to be present (Eugster, 1967; Waser, 1967; Wieland, 1968).

The widely recognized variability in psychoactivity of *A. muscaria* results probably from varying ratios of ibotenic acid and muscimole in the carpophores In spite of appreciable variability between individuals and at different times, certain effects are characteristic: twitching of the limbs, a period of good humor and euphoria, macropsia, occasionally colored visions of the supernatural and illusions of grandeur. Religious overtones frequently occur, and the partaker may become violent, dashing madly about, until exhaustion and deep sleep overtake him.

The genus *Amanita,* of from 50 to 60 species, is cosmopolitan, occurring on all continents except South America and Australia, but this species occupy definitive areas. A number of the species are toxic, and their chemical constitution still poorly understood, appears to be variable.

6. Phenylethylamines

Cactaceae — *Lophophora.* One of the ancient sacred hallucinogens of Mexico, still in use, is the small, grey-green, napiform, spineless cactus 'peyote': *Lophophora williamsii.* It might well be called the 'prototype' of hallucinogens, since it has been one of the most spectacular psychotomimetics known. It was first fully described by the early Spanish medical doctor Francisco Hernández, but many other colonial Spanish croniclers detailed the strange story of peyote. Peyote rites persist in several tribes of northern Mexico. It was used in Texas in 1760, was known among American Indians during the Civil War, but came to public attention in the United States about 1880, when the Kiowas and Comanches elaborated a typical Plains Indian vision-quest ritual around its ceremonial ingestion. The peyote cult, organized as the Native American Church, has gradually spread to many tribes in the United States and Canada and counts 250,000 adherents (La Barre, 1960, 1964; Schultes, 1937a, b, c, 1969-1970). The chloropyll-bearing crown of the cactus, dried into discoidal 'mescal buttons' which are virtually indestructible and can be shipped long distances, is eaten.

The peyote cactus was first botanically described as *Echinocactus williamsii* in 1845. It has frequently been referred to this genus and to *Anhalonium* in the chemical literature. In 1894, it was placed in the monotypic genus *Lophophora.* Its nomenclature and taxonomy are still confused, and *L. williamsii* has more than 25 binomial synonyms, most of them referring to age-forms of the variable crown (Schultes, 1937c).

Lophophora is placed in the tribe *Cereae,* subtribe Echinocactanae, a subtribe of some 28 genera, many of them small or monotypic and which once were included in *Echinocactus* (*Ariocarpus, Astrophytum, Roseocactus,* etc.). It occurs in central Mexico and near the Rio Grande in southern Texas.

More than 30 alkaloids and their amine derivatives have been isolated from *L. williamsii*, belonging mainly to the phenylethylamines and the biogenetically related simple isoquinolines (see Section 4). The phenylethylamine mescaline is exclusively responsible for the visual hallucinations; its derivatives, N-methylmescaline and N-acetylmescaline, are apparently not active. Hordenine, another phenylethylamine, is also present in peyote. Peyonine, a novel phenylethylpyrrole, was recently isolated from the cactus. The pharmacology of this derivative of mescaline or its precursors has not yet been elucidated (Agurell, 1969; Der Marderosian, 1966; Kapadia & Fales, 1968a, b; Kapadia & Highet, 1968; Kapadia, Shah & Zalucky, 1968; Lundström & Agurell, 1967; McLaughlin & Paul, 1965, 1966; Pallares, 1960; Reti, 1950, 1954).

Trichocereus. Several species of the South American genus *Trichocereus* have yielded mescaline: *T. macrogonus, T. pachanoi, T. Terscheckii, T. werdermannianus* (Agurell, 1969). The large columnar *T. pachanoi,* of the dry Andes — called 'San Pedro' in Peru, 'aguacolla' in Ecuador — is employed in magic and folk medicine in northern Peru (Poisson, 1960). Together with another cactus, *Neoraimundia macrostibas* and *Isotoma longiflora, Pedilanthus tithymaloides,* and a species of *Datura*, it is the base of a hallucinogenic drink called 'cimora' (Friedberg, 1964; Gutiérrez-Noriega, 1950). There are some 40 species of *Trichocereus* known from subtropical and temperate South America.

7. Quinolizidines

Leguminosae — *Cytisus* (*Genista*). The hallucinogenic use by Yaqui medicine men in northern Mexico of *Cytisus* (*Genista*) *canariensis,* a shrub native to the Canary Islands, not Mexico, has recently been reported (Fadiman, 1965). It is rich in the toxic alkaloid cystisine (ulexine, baptitoxine, sophorine) which occurs commonly in the Leguminosae (Willaman & Schubert, 1961). About 25 species of *Cytisus,* native to the Atlantic Islands, Europe, and the Mediterranean area, are known, and a number of the species are toxic.

Sophora. A shrub of dry areas of the American Southwest and adjacent Mexico, *Sophora secundiflora* yields the so called 'mescal beans' or 'red beans'. Mexican and Texan Indians formerly employed these beans in the ceremonial Red Bean Dance as an oracular and divinatory medium and for visions in initiation rites (La Barre, 1964; Schultes, 1937a). Its use died out in the United States with the arrival of peyote, a much safer hallucinogen. Mescal beans, which contain cytisine, are capable of causing death by asphyxiation (Henry, 1949; Howard, 1957). Historical reports of the mescal bean go back to 1539, but archaeological remains suggest their ritualistic use earlier than 1000 AD (Campbell, 1958). *Sophora,* with some 50 species, occurs in tropical and warm temperate parts of both hemispheres.

Lythraceae — *Heimia. Heimia salicifolia* has been valued in Mexican folk medicine since earliest times. Known as 'sinicuichi', its leaves are wilted, crushed in water, and the juice set in the sun to ferment The resulting drink is mildly intoxicating. Usually devoid of unpleasant after effects, it induces euphoria characterized by drowsiness, a sense of shrinkage of the surroundings, auditory hallucinations, and a general removal from a sense of reality (Robichaud et al., 1964; Tyler, 1966).

Alkaloids were first reported from *H. salicifolia* in 1958 (Hegnauer & Herfst, 1958). Recent work has isolated and characterized five alkaloids, of which the major psychoactive one appears to be cryogenine (vertine) (Blomster et al., 1964; Douglas et al., 1964). Differing from the usual quinolizidines in having the quinolizidine as part of a larger and complex system of rings, cryogenine has been found only in the Lythraceae. The genus *Heimia* comprises three hardly distinguishable species and ranges from southern United States to Argentina.

Solanaceae — *Atropa.* The belladonna plant, *Atropa belladonna,* was utilized as an hallucinogen in Europe in medieval witches' brews. Its principal active constituent has long been known to be scopolamine, but minor tropane alkaloids are also present (Wagner, 1969). There are four species of *Atropa,* distributed in Europe, the Mediterranean area, and from Central Asia to the Himalayas.

Datura, Methysticodendron. Datura has a long history as an hallucinogenic genus in both hemispheres (Hoffmann, 1968; Lewin, 1964; Safford, 1920, 1921). The genus, comprising some 15 to 20 species, is usually divided into four sections: (a) *Stramonium,* with three species in the two hemispheres; (b) *Dutra,* comprising six species; (c) *Ceratocaulis,* with one Mexican species; (d) *Brugmansia,* South American trees representing possibly six or seven species.

In Asia and the Mediterranean, *D. metel* has been a major narcotic and poison, especially in India. *Datura fastuosa* is smoked for pleasure in Asia and Africa, often along with *Cannabis* and tobacco. Other species were valued in early Europe in witchcraft and as ingredients of sorcerers' potions (Lewin, 1964; Safford, 1920; Wagner, 1969).

In the New World, *Datura* was and is even more widely prized. In ancient and modern Mexico and the American Southwest, 'toloache' (*D. meteloides* or *D. inoxia*) is employed medicinally and as an hallucinogen in divination rites among many tribes. The seeds, foliage, and roots, usually in decoction, are taken. The Indians of parts of northeastern North America made limited use of jimson weed (*D. stramonium*) in adolescence rites (Hoffmann, 1968; Schultes, 1969c,d, 1969-1970).

All of the South American representatives belong to the subgenus Brugmansia, sometimes treated as a distinct genus. They are all arborescent and are native either to the Andean highlands — *D. arborea, D. aurea, D. candida, D. dolichocarpa, D. sanguinea, D. vulcanicola* — or to the warmer lowlands (*D. suaveolens*). Handsome trees, mostly well known in horticultural circles, they all seem to be chromosomally aberrant cultigens, unknown in the wild state. Their classification has long been uncertain. There are six or seven species of this subgenus, although a recent proposal treats them as comprising three species and a number of cultivars.(Schultes, 1960, 1961, 1963a,b, 1965, 1969d).

Some species were of the greatest social and religious importance in ancient Andean cultures. The Chibchas of Colombia, for example, administered potions of *Datura* to wives and slaves of deceased chieftains to induce stupor prior to their being buried alive with the departed master.

The preparation and use of *Datura* differ widely in South America today, but many tribes still employ it for prophecy, divination, and other magico-religious purposes for which the visions are important. The Kamsá of southern Colombia, for example, use several species and numerous named clones, vegetatively propagated and so highly atrophied that they may represent incipient varieties. These monstrous 'races' differ, according to witch doctors, in narcotic strength and are, consequently, used for different purposes (Bristol, 1966, 1969).

What may possibly represent an extreme variant of an indeterminate species of tree-*Datura* has been described and is now known as a distinct genus: *Methysticodendron amesianum.* Native to a high Andean valley of southern Colombia, it is important to the Kamasá and Ingano Indians as an hallucinogen and as medicine (Schultes, 1955).

Chemical work on many of the species of *Datura* has long been carried out, but there is still appreciable variation in results due primarily to failure to insist on authentically vouchered identification of the material analyzed. The principal alkaloids, all tropanes, are hyoscyamine, norhyoscyamine, and scopolamine, present in most of the species; the inactive meteloidine, present in *D. inoxia*; and cusohygrine, found in the roots of several species. It does not contain a tropane ring but biogenetically may be related to the other *Datura*-alkaloids (Henry, 1949; Leete, 1959; Williaman & Schubert, 1961). There are differences in total alkaloid content and in percentage of scopolamine, according to area of cultivation. In Andean plants of *Datura candida,* scopolamine constitutes from 50 to 60 per cent of the total base content, as contrasted to 30 to 34 per cent for the same plants grown in England and Hawaii. Aerial portions of typical *D. candida,* originally from the Colombian Andes but cultivated in England, yielded scopolamine, norscopolamine, atropine, meteloidine, oscine, and noratropine; roots had these alkaloids as well as 3α-6β-ditigloyloxytropane-7β-ol, 3α-tigloyloxytropane, and tropine. Leaves of the same stock grown in Hawaii contained the same spectrum of alkaloids but varied in total content and amount of scopolamine (Bristol, Evans & Lampard, 1969). The leaves and stems of South American material of *Methysticodendron* contained scopolamine up to 80 per cent of the total alkaloid content (Pachter & Hopkinson, 1960).

Significantly, the alkaloidal content in the cultivars of *D. candida* correlate closely with the reports of their relative toxicity by the Indians of Sibundoy, Colombia. Notwithstanding the great age of their hallucinogenic and medicinal uses, *Daturas* are still the subject of much botanical, ethnobotanical, and phytochemical interest.

Hyoscyamus. Henbane, a toxic species of the genus, is *Hyoscyamus niger* and was once widely cultivated in Europe as a narcotic. It entered medieval witches' brews as an hallucinogenic ingredient. The psychoactive effects of henbane are attributed mainly to scopolamine (Wagner, 1969). *Hyoscyamus* comprises about 20 species of Europe, northern Africa, southwestern and central Asia.

Latua. A century ago, a spiny shrub of Chile, now called *Latua pubiflora,* the only member of an endemic genus, was identified as a virulent poison inducing delirium and visual hallucinations. It was employed by local Indians, who knew the shrub as 'latué' or 'arbol de los brujos', to cause permanent insanity (Murillo, 1889). Recent phytochemical studies indicate the presence of atropine and scopolamine (Bodendorf & Kummer, 1962; Silva & Mancinelli, 1959).

Mandragora. The famed mandrake of Europe, *Mandragora officinarum,* owes its renown mainly to its hallucinogenic toxicity. Its active principles are tropane alkaloids, primarily hyoscyamine, scopolamine, and mandragorine (Wagner, 1969). Six species of *Mandragora* are known, native to the region from the Mediterranean to the Himalayas.

9. Tryptamines

Acanthaceae — *Justicia.* The Waikás of the Orinoco headwaters in Venezuela and in northern Amazonian Brazil occasionally dry and pulverize the leaves of *Justicia pectoralis* var. *stenophylla* as an admixture to their Virola-snuff (Schultes, 1966, 1967b; Schultes & Holmstedt, 1968). There are suspicions that this aromatic herb may contain tryptamines (Holmstedt, personal communication). If the preliminary indications can be verified, it will for the first time establish the presence of these indoles in the Acanthaceae. There are more than 300 species of *Justicia* in the tropical and subtropical parts of both hemispheres.

Agaricaceae — *Conocybe, Panaeolus, Psilocybe, Stropharia.* The Spanish conquerors found Mexican Indians practicing religious rites in which mushrooms were ingested as a sacrament permitting them to commune through hallucinations with the spirit world. The Aztecs knew these 'sacred' mushrooms as 'teonanacatl' (food of the gods) (Heim & Wasson, 1959; Safford, 1915).

European persecution drove the cult into hiding in the hinterlands. Notwithstanding the many descriptions in the writings of the early chroniclers, no evidence that the narcotic use of mushrooms had persisted was uncovered until about 30 years ago Botanists had even postulated that teonanacatl was the same plant as peyote: that the discoidal crown of the cactus, when dried, superficially resembled a dried mushroom and that the earlier writers had confused the two or had been deliberately duped by their Aztec informants. Then, during the 1930s, several investigators found an active mushroom cult amongst the Mazatecs in Oaxaca and collected, as the hallucinogenic fungi, *Panaeolus sphinctrinus* and *Stropharia cubensis*. Later and more intensive work during the 1950s brought to over 24 species in at least four genera the number of basidiomycetes employed currently in six or more tribes of Mexican Indians (Guzman, 1959; Heim, 1956a, b, 1957a, b, 1963a; Heim, Hoffmann & Tscherter, 1966; Heim et al., 1967; Schultes, 1939, 1940; Singer, 1958; Wasson, 1958, 1959, 1963; Wasson & Wasson, 1957).

It now appears that the mushroom cults are of great age and were once much more widespread. Archaeological artifacts, now called 'mushroom stones', excavated in great numbers from highland Mayan sites in Guatemala, are dated conservatively at 1000 BC. Consisting of a stem with a human or animal face and crowned with an umbrella-like top, these icons indicate the existence of a sophisticated mushroom cult at least 3000 years ago.

Perhaps the most important species employed in Mexican mushroom rites are *Psilocybe aztecorum, P. caerulescens, P. mexicana, P. zapotecorum,* and *Stropharia cubensis.* (Heim et al., 1967; Heim & Wasson, 1959). All of these have been found to contain a most extraordinary psychactive compound, psilocybine — an hydroxyindole alkylamine with a phosphorylated side chain: 4-phosphoryloxy-N, N-dimethyltryptamine — and sometimes the unstable derivative, psilocine: 4-hydroxy-N, N-dimethyltryptamine. Psilocybine is the only natural indole compound with a phosphoric acid radical known from the plant kingdom, and both psilocybine and psilocine are novel among indoles in having the hydroxy radical substituted in the 4-position. Tryptophan is probably the biogenetic precursor of psilocybine (A. Hoffmann et al., 1958, 1959; Hofmann & Troxler, 1959b; Hofmann & Tscherter, 1960).

These two indoles may occur widely in *Psilocybe* and related genera. One or both have been isolated from *P. baecystis, P. cyanescens, P. fimetaria, P. pelliculosa, P. quebecensis, P. semilanceata, P. semperviva,* and *P. wassonii;* as well as from *Conocybe cyanopus, C. smithii* and a species of *Copelandia* (Benedict et al., 1962, 1967; Heim, Hofmann & Tscherter, 1966; A. Hofmann, 1968; Ola'h & Heim, 1967). The occurrence of 4-substituted tryptamines (psilocybine or psilocine) has been reported from *Panaeolus sphinctrinus,* and this psychoactive mushroom also contains 5-hydroxytryptamine and 5-hydroxytryptophan (Hegnauer, 1962-66; Ola'h, 1969; Tyler & Gröger, 1964). The closely related *Panaeolus campanulatus* does not contain the hallucinogenic constituents (Taylor & Malone, 1960).

Early missionaries in Amazonian Peru reported that the Yurimagua Indians employed an intoxicating beverage made from a 'tree fungus' (Schultes, 1966). Although no modern evidence points to the use of an hallucinogenic fungus in that area, *Psilocybe yungensis* has been suggested as a possible identification of the mushroom (Schultes, 1966).

The principal genera of hallucinogenic mushrooms of Mexico are small but widespread: *Conocybe* is cosmopolitan; *Panaeolus* is cosmopolitan, occurring primarily in Europe, North America, Central America, and temperate Asia; *Psilocybe,* almost cosmopolitan, is distributed in North America, South America, and Asia; and *Stropharia,* likewise almost cosmopolitan, ranges through North America, the West Indies, and Europe.

Leguminosae — *Anadenanthera.* The New World snuff prepared from beans of *Anadenanthera* (*Piptadenia*) *peregrina,* known in the Orinoco basin of Colombia and Venezuela, center of its present use, as 'yopo' or 'nopo', represents probably the 'cohoba' encountered in Hispaniola by Columbus' second voyage in 1496. Von Humboldt, Spruce, and other explorers who mentioned it were all astonished at its hallucinogenic potency (Safford, 1916a; Schultes 1967b; Wassén, 1964, 1967; Wassén & Holmstedt, 1963).

The beans of this medium sized tree, usually roasted, are crushed and mixed with ashes or calcined shells. The powder is ceremonially blown into the nostrils through bamboo tubes or snuffed individually through bird-bone tubes. The intoxication is marked by fury, followed by an hallucinogenic trance and eventual stupors (Granier-Doyeux, 1965).

Five indoles have been isolated from *A. peregrina,* chief of which are N,N-dimethyltryptamine and bufotenine (5-hydroxy-N,N-dimethyltryptamine (Fish et al., 1955; Holmstedt & Lindgren, 1967). The beans contain as their main constituent N,N-dimethyltryptamine or bufotenine. Other indoles found in this species are 5-methoxy-N,N-dimethyltryptamine, N-monomethyltryptamine and 5-methoxy-N-monomethyltryptamine.

Indirect evidence suggests that another species, *A. colubrina,* might formerly have been the source of the narcotic snuffs known in southern Peru and Bolivia as 'vilca' or 'huilca' and in northern Argentina as 'cebil' (Altschul, 1967). Since this species is closely related to the more northern *A. peregrina* and its chemical constituents are very similar, *A. colubrina* may well have been valued aboriginally as an hallucinogen.

Anadenanthera comprises only the two species discussed above. Native to South America, they are distinguished from the closely allied genus *Piptadenia* both morphologically and chemically (Altschul, 1964).

Mimosa. The allied genus *Mimosa* likewise yields a psychotomimetic, 'vinho de jurema'. An infusion of the roots of *Mimosa hostilis* forms the center of the ancient Yurema cult of the Karirí, Pankarurú, and other Indians of Pernambuco State, Brazil (Schultes, 1965, 1966). The drink, said to induce glorious visions of the spirit world, was reported to contain an alkaloid called nigerine, now known to be synonymous with N,N-dimethyltryptamine, the active principle (Gonçalves de Lima, 1946; Pachter & Hopkinson, 1960).

The genus *Mimosa* comprises about 500 tropical or subtropical herbs and small shrubs, mostly American but a few native to Africa and Asia. It is closely related to *Anadenanthera* and *Piptadenia.*

Malpighiaceae — *Banisteriopsis.* One of the numerous admixtures of the 'ayahuasca-caapi-yajé' drink prepared basically from bark of *Banisteriopsis caapi* or *B. inebrians* (which contain β-carboline bases) is the leaf of *B. rusbyana,* known in the western Amazon of Colombia and Ecuador as 'oco-yajé'. The natives add the leaf to heighten and lengthen the visions. Recent examination indicates that *B.rusbyana* has in its leaves and stems, to the exclusion of the harmala alkaloids characteristic of the other two narcotically utilized species, N,N-dimethyltryptamine and traces of other tryptamines (N_3-methyltryptamine; 5-methoxy-N,N-dimethyltryptamine; and N,-methyltetrahydro-β-carboline) (Agurell, Holmstedt & Lindgren, 1968; Der Marderosian, Pinkley & Dobbins, 1968; Poisson, 1965). Tryptamines have apparently not hitherto been reported from the Malpighiaceae.

Myristicaceae — *Virola.* Hallucinogenic snuffs are prepared in northwestern Brazil and adjacent Colombia and Venezuela from the reddish bark resin of *Virola,* a genus of 60 to 70 trees of Central and South America. The species employed have only recently been identified as *V. calophylla* and *V. calophylloidea* in Colombia and *V. theiodora* in Brazil (Schultes, 1954b, c; Schultes & Homstedt, 1968; Seitz, 1967). The most intense use of this snuff, called 'yakee', 'parica', 'epena', and 'nyakwana', centers among the Waikás of Brazil and Venezuela. In Colombia, only witch doctors employ it, but in Brazil the intoxicant is taken by all adult males, either individually at any time or ritually in excess at endocannibalistic ceremonies amongst the Waikás. The resin, which is boiled, dried, pulverized, and occasionally mixed with powdered leaves of a *Justicia* and bark-ashes of *Theobroma subinacanum* or *Elizabetha princeps*, acts rapidly and violently. Effects include excitement, numbness of the limbs, twitching of facial muscles, nausea, hallucinations, and finally a deep sleep; macropsia is frequent and enters into Waiká beliefs about the spirits resident in the drug.

Contemporary investigations indicate that the snuff prepared from *V. theiodora* contains normally up to 8 per cent 5-methoxy-N,N-dimethyltryptamine, with lesser amounts of N,N-dimethyltryptamine (Agurell, Holmstedt, Lindgren & Schultes, 1969; Holmstedt, 1965). There is appreciable variation in alkaloid concentration in different parts (leaves, bark, root) of *V. theiodora,* but the content in the bark resin may reach as high as 11 per cent. Two new β-carbolines have likewise been found in *V. theiodora* (Agurell, Holmstedt, Lindgren & Schultes, 1968).

Of other species of *Virola* investigated, *V. rufula* contains substantial amounts of tryptamines and *V. calophylla,* one of the species employed in the preparation of snuff in Colombia, contains high amounts of alkaloids apparently in the leaves alone. *V. multinervia* and *V. venosa* are almost devoid of alkaloids (Agurell, Holmstedt, Lindgren & Schultes, 1969).

The Witotos, Boras, and Muinanes of Amazonian Colombia utilise the resin of a *Virola,* possibly *V. theiodora,* orally as an hallucinogen. Small pellets of the boiled resin are rolled in a 'salt' left upon evaporation of the filtrate of bark ashes of *Gustavia poeppigiana* and ingested to bring on a rapid intoxication, during which the witch doctors see and speak with 'the little people' (Schultes, 1969b). There are suggestions that Venezuelan Indians may smoke *V. sebifera* as an intoxicant.

Rubiaceae — *Psychotria.* Among the sundry admixtures employed to 'strengthen' and 'lengthen' the effects of the hallucinogenic drink prepared from *Banisteriopsis caapi* and *B. inebrians* in the western Amazon, one of the most commonly added are leaves of *Psychotria* (Schultes, 1967a). One species used in Ecuador and Peru, *P. viridis* (reported through a misidentification as *P. psychotriaefolia* (Schultes, 1966, 1969a)), has recently been shown to contain N,N-dimethyltryptamine (Der Marderosian, personal communication). The same species and another not yet specifically identified are similarly used in Acre Territory, Brazil (Prance, in press). Tryptamines have apparently not hitherto been reported from the Rubiaceae. The genus *Psychotria* comprises more than 700 species of the warmer parts of both hemispheres, many of which have important roles in folk medicine or are poisons.

HALLUCINOGENS OF UNCERTAIN USE OR CHEMICAL COMPOSITION

Sundry plants known to possess psychoactive constituents are doubtfully employed as hallucinogens. Others are known to be used for their psychotomimetic properties, but the chemical principles responsible for the effects are of uncertain or undetermined structure.

Lycoperdaceae — *Lycoperdon.* Puffballs, *Lycoperdon marginatum* and *L. mixtecorum,* have recently been reported as hallucinogens utilized by the Mixtecs of Oaxaca in Mexico at 6000 feet altitude or higher (Heim et al., 1967). There are more than 100 species of *Lycoperdon,* native mostly to the temperate zone in moss-covered forests.

The Mixtecs call *Lycoperdon mixtecorum* 'gi'-i''wa' (fungus of first quality) and *L. marginatum,* which has a strong odor of excrement, 'gi'-i-sa-wa' (fungus of second quality). These two hallucinogens do not appear to occupy the place as divinatory agents that the mushrooms hold among the neighboring Mazatecs.

The more active species, *Lycoperdon mixtecorum,* causes a state of half-sleep one-half hour after ingestion of one or two specimens. Voices and echoes are heard, and voices are said to respond to questions posed to them. The effects of the puffballs differ strongly from those of the hallucinogenic mushrooms: they may not induce visions, although definite auditory hallucinations do accompany the intoxication. There is as yet no phytochemical basis on which to explain the intoxication from these two gastromycetes.

Araceae — *Acorus.* There is some evidence that Indians of northern Canada chewed the root of *Acorus calamus* — flag root, rat root, sweet calomel — for its medicinal and stimulant properties. In excessive doses this root is known to induce strong visual hallucinations (Hoffer & Osmond, 1967; Sharma et al., 1961). The hallucinogenic principle is reported to be α-asarone and β-asarone (Hoffer & Osmond, 1967). There are two species of *Acorus* occurring in the north temperate zone and warmer parts of both hemispheres.

Homalomena. Natives of Papua are reported to eat the leaves of ereriba, a species of *Homalomena,* together with the leaves and bark of *Galbulimima belgraviana,* as a narcotic. The effects are a violent and crazed condition leading to deep sleep, during which the partakers see and dream about the men or animals that they are supposed to kill (Barrau, 1957, 1958; Hamilton, 1960). It is not yet clear what, if any, hallucinogenic principle may be present in this aroid. Some 140 species of *Homalomena,* native to tropical Asia and South America are known.

Amaryllidaceae — *Pancratium.* The Bushmen of Dobe, Botswana, consider *Pancratium trianthum,* a bulbous perennial known locally as 'kwashi', to be psychoactive (Schultes, 1969-1970). Rubbing the bulb over incisions on the head is said to induce visual hallucinations. Nothing is known of possible psychotomimetic constituents. Other species of *Pancratium,* a genus of some 15 species, mainly of Asia and Africa, possess toxic principles, chiefly alkaloids. Although some species are employed in folk medicine, several are potent cardiac poisons.

Zingiberaceae — *Kaempferia.* Vague reports indicate that in New Guinea, *Kaempferia Galanga,* known as 'maraba', is employed as an hallucinogen (Barrau, 1962; Hamilton, 1960), but phytochemical corroboration is lacking. The rhizome of 'galanga', containing essential oils, is highly prized as a condiment and medicine in tropical Asia. There are some 70 species of *Kaempferia* distributed in tropical Africa, India to southern China, and western Malaysia.

Moraceae — *Maquira.* An Amazon jungle tree, *Maquira sclerophylla,* represents one of the most poorly understood hallucinogens. The fruits reputedly are the source of an intoxicating snuff employed formerly by Indians of the Pariana region of central Amazonia (Schultes, 1960, 1961, 1963a, b, 1965, 1966, 1969d). Nothing is known of their chemical constituents. Two species of *Maquira* — once called *Olmedioperebea* — have been described from the Amazon.

Aizoaceae — *Mesembryanthemum.* More than 225 years ago, the Hottentots of South Africa were reported using a narcotic called 'kanna' or 'channa'. At the present time, this name applies to sundry species of *Mesembryanthemum* (*Sceletium*), especially to *M. tortuosum,* but there is no evidence that these are employed hallucinogenically. Other plants — *Sclerocarya caffra* of the Anacardiaceae and *Cannabis* — have been suggested as possible identifications (Lewin, 1964; Schultes, 1967a, 1969-1970; Tyler, 1966).

Several species of *Mesembryanthemum* known to cause a state of torpor when ingested have yielded alkaloids; mesembrine and mesembrenine Both have a nucleus related to the crinane nucleus in certain amaryllidaceous alkaloids but differ in having an open ring. There are about 1000 species of *Mesembryanthemum,* sensu lato, in the xerophytic parts of South Africa. About two dozen species have been split off into a group often recognized as a distinct genus, *Sceletium.*

Himantandraceae — *Galbulimima.* In Papua, the leaves and bark of 'agara', *Galbulimima belgraveana,* are taken with the leaves of a species of *Homalomena* to induce a violent intoxication that progresses into a sleep in which visions and dreams are experienced (Barrau, 1957, 1958; Hamilton, 1960). Several isoquinoline alkaloids have been isolated from this plant, but the specific pharmacology of the constituents is not clear. Two or three species of *Galbulimima* occur in eastern Malaysia and northeastern Australia

Gomortegaceae — *Gomortega. Gomortega keule,* an endemic of Chile, where it has the Mapuche Indian names 'keule' or 'hualhual', may once have been employed as a narcotic (Mariani Ramirez, 1965; Mechoulam & Gaoni, 1965). Its fruits are intoxicating, especially when fresh, due possibly to an essential oil. There is only this one species in the Gomortegaceae.

Leguminosae — *Erythrina.* The reddish beans of *Erythrina* may have been valued as hallucinogens in Mexico. Resembling seeds of *Sophora secundiflora,* they are frequently sold in modern Mexican herb markets under the name 'colorines' (Safford, 1916b; Schultes, 1937b, 1969c, 1969-1970). Several species contain indole or isoquinoline derivatives and could be hallucinogenic. The genus occurs in the tropics and subtropics of both hemispheres and comprises some 100 species.

Rhynchosia. The ancient Mexicans may have valued several species of *Rhynchosia* as a narcotic. Modern Oaxacan Indians refer to the toxic seeds of *R. pyramidalis* and *R. longeracemosa* by the same name, 'piule', that they apply to the seeds of hallucinogenic morning glories. The black and red *Rhynchosia* beans, pictured together with mushrooms, have been identified on Aztec paintings, thus suggesting hallucinogenic use (Schultes, 1937b, 1965, 1969c, 1969-1970). An as yet uncharacterized alkaloid has been isolated from this genus, which comprises some 300 species of the tropics and subtropics, especially of Africa and America

Malpighiaceae — *Tetrapteris.* The Makú Indians in the northwesternmost sector of the Brazilian Amazon prepare a narcotic drink from the bark of *Tetrapteris methystica.* A cold-water infusion with no admixtures has a yellowish hue and induces an intoxication with visual hallucinations very similar to that caused by drinks prepared from species of the related genus *Banisteriopsis* (Schultes, 1954a).

No chemical studies have been made of this species of *Tetrapteris,* but, since it is close to *Banisteriopsis,* it is not improbable that β-carbolines are the active constituents. *Tetrapteris* comprises some 80 species distributed from Mexico to tropical South America and in the West Indies.

Coriariaceae — *Coriaria.* Long recognized in the Andes as dangerously toxic to animals, *Coriaria thymifolia* has recently been reported as hallucinogenic, giving the sensation of flight. The fruits, reputedly containing catecholic derivatives, are eaten for inebriation in Ecuador, where the plant is called 'shanshi' (Naranjo, 1969; Naranjo & Naranjo, 1961).

Four toxic picrotoxine-like sesquiterpenes have been isolated from the Coriariaceae: coriamyrtine, coriatine, tutine, and pseudotutine (Hegnauer, 1962-66). This genus, the only one in the family, has some 15 species distributed in Eurasia, New Zealand, and highland tropical America.

Cactaceae — *Ariocarpus.* The Tarahumare Indians of northern Mexico employ *Ariocarpus fissuratus,* called 'sunami' and 'peyote cimarrón', as a narcotic, asserting that it is stronger than true peyote (*Lophophora)* (Schultes, 1967a, 1969-1970). Anhalonine has been isolated from an indeterminate species of *Ariocarpus.* There are five species known in this genus, all Mexican (Agurell, 1969; Der Marderosian, 1967a).

Epithelantha. The Tarahumare likewise use *Epithelantha micromeris* as a narcotic (Schultes, 1963a). Chemical studies apparently have not been carried out on representatives of this genus of three species of southwestern United States and Mexico.

Pachycereus. Another cactus utilized as a narcotic by the Tarahumare is the gigantic *Pachycereus pecten-aboriginum,* which they call 'cawe'. Carnegine has been reported from this species (as *Cereus pecten-aboriginum*) (Agurell, 1969). Another species, *P. marginatus,* is said to contain pilocereine (Agurell, 1969). There are five species of *Pachycereus,* all native to Mexico.

Ericaceae — *Pernettya. Pernettya furiens,* known in Chile as 'huedhued' or 'hierba loca', is toxic. When consumed in quantity, the fruits induce mental confusion and madness or permanent insanity and exercise a narcotic effect similar to that of *Datura* (Mariani Ramirez, 1965). This species has apparently not been chemically investigated. Its activity may be due to andromedotoxine, a resinoid, or to arbutin, a glycoside, or hydroquinone — both rather widely distributed in the family. *P. parvifolia*, called 'taglli' in Ecuador, is noted as a toxic plant containing andromedotoxine and the fruit of which, when ingested, causes hallucinations and other psychic and motor alterations (Chavez et al., 1967; Naranjo, 1969). Some 25 species of *Pernettya* are known from Tasmania, New Zealand, the highlands from Mexico to Chile, the Galapagos and Falkland Islands

Desfontainiaceae — *Desfontainia.* It is reported that the leaves of *Desfontainia spinosa* var. *hookeri,* are employed in southern Chile as a narcotic as well as medicinally (Mariani Ramirez, 1965). Chemical investigation is this anomalous plant have apparently not been carried out. This genus of two or three Andean species comprises the only genus in the family, which appears to be related to the Loganiaceae and which is sometimes placed in the Potaliaceae.

Apocynaceae — *Prestonia.* The source of the hallucinogenic 'yajé' of the western Amazon has been reported as *Prestonia* (*Haemadictyon*) *amazonica,* an identification based on misinterpretation of field data and guesswork. Although well established in botanical and chemical literature, recent evaluation of the evidence seriously discredits this suggestion (Schultes & Raffauf, 1960). A recent report of N,N-dimethyltryptamine in *P. amazonica* (Hochstein & Paradies, 1957) was based upon an erroneous identification without voucher specimens of an aqueous extract of the leaves of a vine which may have been *Banisteriopsis Rusbyana.*

Labiatae — *Coleus, Salvia.* In southern Mexico, crushed leaves of *Salvia divinorum,* known in Oaxaca as 'hierba de la Virgen' or 'hierba de la Pastora', are valued by the Mazatecs in divinatory rites when other more potent hallucinogens are unavailable (Epling & Jativa-M., 1962). Although investigators have experimentally substantiated the psychotomimetic effects, a toxic principle is still to be isolated from the plant (Wasson, 1962a,b). It has been suggested that *S. divinorum* represents the hallucinogenic 'pipilzintzintli' of the ancient Aztecs (Wasson, 1962a). There are some 700 species of *Salvia* in the temperate and tropical parts of both hemispheres, but no other species seems to have been reported as an hallucinogen.

The leaves of two other mints, *Coleus pumila* and *C. blumei,* both native to southeast Asia, are similarly employed by the Mazatecs (Wasson, 1962a). Chemical studies of these two species, at least on the basis of the material growing in southern Mexico, have not been done, and a psychoactive principle is not known in this genus of some 150 species of the Old World tropics.

Solanaceae — *Brunfelsia.* Several species of *Brunfelsia* have been reported used narcotically in the western Amazon. Several vernacular names have long suggested that the intoxicating properties were valued (Schultes, 1967a). *B. tastevinii* is reputed utilized by the Kachinauas of the Brazilian Amazon to prepare an hallucinogenic drink, but this report needs confirmation (Benoist, 1928). Containing what appear to be tropanes of undetermined structure and the coumarin compound scopoletin, this genus undoubtedly has psychoactive properties. *Brunfelsia* is a tropical American genus of some 25 species, and is somewhat intermediate between the Solanaceae and Scrophulariaceae.

Solanaceae — *Iochroma.* In the Andes of southern Colombia, *Iochroma fuchsioides* is employed as a divinatory narcotic in difficult cases. Known as 'borrachera', it is grown especially in gardens of medicine-men. No chemical examination has as yet been conducted. There are some 25 species of Iochrome native to South America.

Campanulaceae — *Lobelia. Lobelia tupa,* a tall, polymorphic herb of the Andean highlands known as 'tupa' or 'tabaco del diablo', is a widely recognized poison. Chilean peasants are said to employ the juice to relieve toothache, and while the Mapuches of Chile reputedly smoke the leaves for their narcotic effect, there is yet no certainty that this effect is hallucinogenic (Mariani Ramirez, 1965; Naranjo, 1969).

The leaves of *L. tupa* contain the piperidine alkaloid lobeline and the diketo- and dihydroxy-derivatives, lobelamidine and norlobelamidine (Kaczmarek & Steinegger, 1958, 1959).

There are some 350 to 400 cosmopolitan species of *Lobelia,* mostly tropical and subtropical, especially in the Americas. It is usually classified with several other large genera as a subfamily, Lobeliodieae, of the Campanulaceae, but the subfamily may sometimes be treated as a distinct family, Lobeliaceae.

Compositae — *Calea.* A common Mexican shrub, *Calea zacatechichi,* belonging to a tropical American genus of about 100 species, represents one of the most recently discovered hallucinogens. The Chontal Indians of Oaxaca take the leaves in infusion for divination, calling them 'thle-pela-kano' (leaf of god) and believing them to clarify the senses (MacDougall, 1968). Although the plant has long been used in folk medicine, few reliable chemical studies appear to have been carried out (Schultes, 1969c). Preliminary investigations have indicated the presence of a possible new alkaloid (Holmstedt, personal communication).

BIBLIOGRAPHY

Agurell, S., 1969, in Lloydia, 32:206-216.

Agurell, S., B. Holmstedt & J.-E. Lindgren, 1968, in Am.J.Pharm., 140:1-4.

Agurell, S., B. Holmstedt, J.-E. Lindgren & R. E. Schultes, 1968, in Biochem. Pharmacol., 17:2487-2488.

Agurell, S., B. Holmstedt, J.-E. Lindgren & R. E. Schultes, 1969, in Acta Chem.Scand., 23:903-916.

Altschul, S. von R., 1964, in Contrib. Gray Herb., Harvard Univ., 193:1-65.

Altschul, S. von R., 1967, pp. 307-314 in D. Efron (ed.), Ethnopharmacologic search for psychoactive drugs. Washington, D.C.: US Pub.Health Serv., Publ.No.1645.

Barrau, J., 1957, in J.Agr.Trop.Bot.Appl., 4:348-349.

Barrau, J., 1958, in J.Agr.Trop.Bot.Appl., 5:377-378.

Barrau, J., 1962, in J.Agr.Trop.Bot.Appl., 9:245-249.

Benedict, R. G., L. R. Brady, A. H. Smith & V. E. Tyler, Jr., 1962, in Lloydia, 25:156-159.

Benedict, R. G., V. E. Tyler, Jr. & R. Watling, 1967, in Lloydia, 30:150-157.

Benoist, R., 1928, in Bull.Soc.Bot.Fr., 75:295.

Blomster, R. N., A. E. Schwarting & J. M. Bobbitt, 1964, in Lloydia, 27:15-24.

Bodendorf, K. & H. Kummer, 1962, in Pharm.Zentralb., 101:620-622.

Brekhman, I. I. & Y. A. Sam, 1967, p. 415 in D. Efron (ed.), Ethnopharmacologic search for psychoactive drugs. Washington, D.C.: US Pub.Health Serv., Publ. No. 1645.

Bristol, M. L., 1966, in Bot.Mus.Leafl., Harvard Univ., 21:229-248.

Bristol, M. L., 1969, in Bot.Mus.Leafl., Harvard Univ., 22:165-227.

Bristol, M. L., W. C. Evans & J. F. Lampard, 1969, in Lloydia, 32:123-130.

Bunge, A., 1847, in Mem.Sav.Etr.Petersb., 7:438.

Campbell, T. N., 1958, in Am.Anthrop., 60:156-160.

Chavez, L., E. de Naranjo & P. Naranjo, 1967, in Cienc.Natur., 9:16.

Chen, A. L. & K. K. Chen, 1939, in Quart.J.Pharm.Pharmacol., 12:30-38.

Cook, W. B. & W. E. Keeland, 1962, in J.Org.Chem., 27:1061-1062.

Der Marderosian, A. H., 1965, in Taxon., 14:234-240.

Der Marderosian, A. H., 1966, in Am.J.Pharm., 138:204-212.

Der Marderosian, A. H., 1967a, in Lloydia, 30:23-38.

Der Marderosian, A. H., 1967b, in Am.J.Pharm., 139:19-26.

Der Marderosian, A. H., H. V. Pinkley & M. F. Dobbins, 1968, in Am.J.Pharm 140:137-147.

Der Marderosian, A. H. & H. W. Youngken, Jr., 1966, in Lloydia, 29:35-42.

Deulofeu, V., 1967, pp. 393-402 in D. Efron (ed.), Ethnopharmacologic search for psychoactive drugs. Washington, D.C.: US Pub.Health Serv., Publ. No. 1645.

Douglas, B., J. L. Kirkpatrick, R. F. Raffauf, O. Ribeiro & J. A. Weisbach, 1964, in Lloydia, 27:25-31.

Epling, C. & C. D. Jativa-N., 1962, in Bot.Mus.Leafl., Harvard Univ., 20:75-76.

Eugster, C. H., 1967, pp. 416-418 in D. Efron (ed.), Ethnopharmacologic search for psychoactive drugs. Washington, D.C.: US Pub.Health Serv., Publ. No. 1645.

Fadiman, J., 1965, in Econ.Bot., 19:383.
Farnsworth, N. R., 1968, in Science, 162:1086-1092.
Farnsworth, N. R., 1969, pp.367-399 in J. E. Gunckel (ed.), Current topics in plant science. New York: Academic Press.
Fish, M. S., N. M. Johnson & E. C. Horning, 1955, in J.Am.Chem.Soc., 77: 5892-5895.
Friedberg, C., 1964, in Sixth Int.Congr.Anthrop.Ethnol.Sci., 2: pt. 2, 21-26.
Friedberg, C., 1965, in J.Agr.Trop.Bot.Appl., 12:403-437, 550-594, 729-780.
Genest, K., W. B. Rice & C. G. Farmilo, 1965, in Proc.Can.Soc.Forensic Sci., 4:167-186.
Gonçalves de Lima, O., 1946, in Arq.Inst.Pesqui.Agron.Recife, 4:45-80.
Granier-Doyeux, M., 1965, in Bull.Narcotics, 17:29-38.
Gutiérrez-Noriega, C., 1950, in América Indig., 10:215-220.
Guzmán, H. G., 1959, in Bol.Soc.Bot.Mex., 24:14-34.
Hamilton, L., 1960, in Papua New Guinea Sci.Soc.Trans., 1:16-18.
Hegnauer, R., 1962, 1963, 1964, 1966, Chemotaxonomie der Pflanzen, 1, 2, 3, 4. Basle: Birkhauser Verlag.
Hegnauer, R. & A. Herfst, 1958, in Pharm.Weekbl., 93-849.
Heim, R., 1956a, in C.R.Acad.Sci., Ser.D, 242:965-968.
Heim, R., 1956b, in C.R.Acad.Sci., Ser.D, 242-1389-1395.
Heim, R., 1957a, in C.R.Acad.Sci., Ser.D, 244-659-700.
Heim, R., 1957b, in Rev.Mycol., 22:58-79, 183-198.
Heim, R., 1963a, in C.R.Acad.Sci., Ser.D, 254:788-791.
Heim, R., 1963b, Les champignons toxiques et hallucinogènes. Paris: N. Boutée & Cie.
Heim, R., A. Hofmann & H. Tacherter, 1966, in C.R.Acad.Sci., Ser.D, 262:51.
Heim, R. et al., 1967, Nouvelles investigations sur les champignons hallucinogènes. Paris: Editions du Musée National d'Histoire Naturelle.
Heim, R. & R. G. Wasson, 1959, Les champignons hallucinogènes du Mexique. Paris: Editions du Musée National d'Histoire Naturelle.
Henry, T. A., 1949, The plant alkaloids. Philadelphia, Pa.: Blakiston.
Hochstein, F. A. & A. M. Paradies, 1957, in J.Am.Chem.Soc., 79:5735-5736.
Hoffer, A. & H. Osmond, 1967, The hallucinogens. New York: Academic Press.
Hoffmann, R. M., 1968, Datura: its use among Indian tribes of Southwestern North America. Unpubl.ms., Bot.Mus.Harvard Univ., Cambridge, Mass.
Hofmann, A., 1961a, in J.Exp.Med.Sci., 5:31-51.
Hofmann, A., 1961b, in Planta Med., 9:354-367.
Hofmann, A., 1963a, in Indian J.Pharm., 25:245-256.
Hofmann, A., 1963b, in Bot.Mus.Leafl., Harvard Univ., 20:194-212.
Hofmann, A., 1964, in Planta Med., 12:341-352.
Hofmann, A., 1966, in Colloq.Int.Centr.Nat.Rech.Sci., No.144:223-241.
Hofmann, A., 1967, in Therapiewoche, 17:1739.
Hofmann, A., 1968, pp.169-235 in A.Burger (ed.), Chemical constitution and pharmacodynamic action, 2. New York: Dekker Publ.
Hofmann, A. & A. Cerletti, 1961, in Deut.Med.Wochenschr., 86: 885-894.
Hofmann, A., A. Frey, H. Ott, T. Petrzilka & F. Troxler, 1958a, in Experientia, 14:397-401.
Hofmann, A., R. Heim, A. Brack & H. Kobel, 1958b, in Experientia, 14:107-109.
Hofmann, A., R. Heim, A. Brack, H. Kobel, A. Frey, H. Ott, T. Petrzilka & F. Troxler, 1959a, in Helv.Chim.Acta, 42:1557-1572.

Hofmann, A. & F. Troxler, 1959b, in Experientia, 15:101-104.
Hofmann, A. & H. Tscherter, 1960, in Experientia, 16:414-416.
Holmstedt, B., 1965, in Arch.Int.Pharmacodyn.Ther., 156:285-304.
Holmstedt, B. & J.-E. Lindgren, 1967, pp.339-373 in D. Efron (ed.), Ethnopharmacologic search for psychoactive drugs. Washington, D.C.: US Pub. Health Serv., Publ. No. 1645.
Howard, J. H., 1957, in Am.Anthrop., 59:75-87.
Hylin, J. W. & D. P. Watson, 1965, in Science, 148:499-500.
Kaczmarek, F. & E. Steinegger, 1958, in Pharm.Acta Helv., 33:257, 852.
Kaczmarek, F. & E. Steinegger, 1959, in Pharm.Acta Helv., 34:413.
Kapadia, G. J. & H. M. Fales, 1968a, in Chem.Commun.:1688-1689.
Kapadia, G. J. & H. M. Fales, 1968b, in J.Pharm.Sci., 57:2017-2018.
Kapadia, G. J. & R. J. Highet, 1968, in J.Pharm.Sci., 57:191-192.
Kapadia, G. J., N. J. Shah and T. B. Zalucky, 1968, in J.Pharm.Sci., 57:254-262.
Korte, F. & H. Sieper, 1965, pp.15-30 in Ciba Found. Study Group No. 21, Hashish: its chemistry and pharmacology. London: Churchill.
La Barre, W., 1960, in Curr.Anthrop., 1:45-60
La Barre, W., 1964, The peyote cult (enlarged ed.). Hamden, Conn.: Shoe String Press.
Leete, E., 1959, pp.48-56 in A. G. Avery, S. Satina & J. Rietsema (eds.), Blakeslee: the genus datura. New York: Ronald Press.
Lewin, L., 1964, Phantastica, narcotic and stimulating drugs. London: Routledge & Kegan Paul.
Lundstrom, J. & S. Agurell, 1967, in J.Chromatogr., 30:271-272.
MacDougall, T., 1960, in Bol.Centro Inv.Antropol.Mex., 6:6.
MacDougall, T., 1968, in Gard.J., 18:105.
Mariani Ramírez, C., 1965, pp. 329-371 in A. Bello (ed.), Témas de Hipnosis. Santiago de Chile.
McLaughlin, J. L. & A. G. Paul, 1965, in J.Pharm.Sci., 54:661.
McLaughlin, J. L. & A. G. Paul, 1966, in Lloydia, 29:315-327.
Mechoulam, R., P. Braun & Y. Gaoni, 1967, in J.Am.Chem.Soc., 89:4552-4554.
Mechoulam, R. & Y. Gaoni, 1965, in J.Am.Chem.Soc., 87:3273-3275.
Mors, W. B. & P. Zaltzman, 1954, in Bol.Inst.Quim.Agr., No.34:17-27.
Murillo, A., 1889, p.152 in A. Roger & F. Chernovicz (eds.), Plantes médicinales du Chili. Paris: Imprimerie de Lagny.
Naranjo, P., 1969, in Terapia, 24:5-63.
Naranjo, P. & E. de Naranjo, 1961 in Arch.Crimin.Neuro-Psiqu.Discpl. Conexas, 9:600.
O'Connell, F. D. & E. V. Lynn, 1953, in J.Am.Pharm.Assoc., 42:753-754.
Ola'h, G.-M., 1969, in Rev.Mycol., 33:284-290.
Ola'h, G.-M. & R. Heim, 1967, in C.R.Acad.Sci., Ser.D, 264:1601-1604.
Osmond, H., 1955, in J.Ment.Sci., 101:526-537.
Pachter, I. J. & A. F. Hopkinson, 1960, in J.Am.Pharm.Assoc.Sci.Ed., 49:621-622.
Pallares, E. S., 1960, in Cactac.Sucul.Mex., 5:35-43.
Pérezamador, M. C. & J. Herrán, 1960, in Tetrahedron Lett., 30.
Poisson, J., 1960, in Ann.Pharm.Fr., 18:764-765.
Poisson, J., 1965, in Ann.Pharm.Fr., 23:241-244.

Pope, H. G., Jr., 1969, in Econ.Bot., 23:174-184.
Porter, D. M., 1962, The taxonomic and economic uses of Peganum (Zygophyllaceae). Unpublished ms., Bot.Mus.Harvard Univ., Cambridge, Mass.
Prance, G. T., Econ.Bot. (in press).
Reti, L., 1950, in Fortschr.Chem.Org.Naturst., 6:242-289.
Reti, L., 1954, pp.7-28 in R. H. F. Manske & H. L. Holmes (eds.), The alkaloids, 4. New York: Academic Press.
Robichaud, R. C., M. H. Malone & A. E. Schwarting, 1964, in Arch.Int. Pharmacodyn.Ther., 150:220-232.
Safford, W. E., 1915, in J.Hered., 6:291-311.
Safford, W. E., 1916a, in J.Wash.Acad.Sci., 6:547-562.
Safford, W. E., 1916b, in Smithson.Inst.Ann.Rep. 1916: 387-424.
Safford, W. E., 1920, in Smithson.Inst.Ann.Rep. 1920: 537-567.
Safford, W. E., 1921, in J.Wash.Acad.Sci., 11:173-189.
Satina, S. & A. G. Avery, 1959, pp.16-47 in A. G. Avery, S. Satina & J. Rietsema (eds.), Blakeslee: the genus datura. New York: Ronald Press.
Schultes, R. E., 1937a, in Bot.Mus.Leafl., Harvard Univ., 4:129-152.
Schultes, R. E., 1937b, in Bot.Mus.Leafl., Harvard Univ., 5:61-88.
Schultes, R. E., 1937c, Peyote (Lophophora Williamsii (Lemaire) Coulter) and its uses. Unpubl. thesis, Harvard Univ., Cambridge, Mass.
Schultes, R. E., 1939, in Bot.Mus.Leafl., Harvard Univ., 7:37-54.
Schultes, R. E., 1940, in Am.Anthrop., 42:429-443.
Schultes, R. E., 1941, A contribution to our knowledge of Rivea corymbosa, the narcotic Ololiuqui of the Aztecs. Cambridge, Mass.: Bot.Mus.Harvard Univ.
Schultes, R. E., 1954a, in Bot.Mus.Leafl., Harvard Univ., 16:202-205.
Schultes, R. E., 1954b, in Bot.Mus.Leafl., Harvard Univ., 16:241-260.
Schultes, R. E., 1954c, in J.Agric.Trop.Bot.Appl., 1:298-311.
Schultes, R. E., 1955, in Bot.Mus.Leafl., Harvard Univ., 17:1-11.
Schultes, R. E., 1957, in Bot.Mus.Leafl., Harvard Univ., 18:1-56.
Schultes, R. E., 1960, in Pharm.Sci., Third Lect.Ser., 141-167.
Schultes, R. E., 1961, in Texas J.Pharm., 2:141-167, 168-185.
Schultes, R. E., 1963a, in Harvard Rev., 1:18-32.
Schultes, R. E., 1963b, in Psyched.Rev., 1:145-166.
Schultes, R. E., 1964, in Taxon., 13:65-66.
Schultes, R. E., 1965, in Planta Med., 13:125-157.
Schultes, R. E., 1966, in Lloydia, 29:293-308.
Schultes, R. E., 1967a, pp.33-57 in D. Efron (ed.), Ethnopharmacologic search for psychoactive drugs. Washington, D.C.: US Pub.Health Serv., Publ. No. 1645.
Schultes, R. E., 1967b, pp.291-306 in D. Efron (ed.), Ethnopharmacologic search for psychoactive drugs. Washington, D.C.: US Pub.Health Serv., Publ. No. 1645.
Schultes, R. E., 1968a, in Rhodora, 70:313-339.
Schultes, R. E., 1968b, in Rev.Cienc.Cultura, 20:37-49.
Schultes, R. E., 1969a, in Bot.Mus.Leafl., Harvard Univ., 22:133-164.
Schultes, R. E., 1969b, in Bot.Mus.Leafl., Harvard Univ., 22:229-240.
Schultes, R. E., 1969c, in Science, 163:245-254.
Schultes, R. E., 1969d, pp.336-354 in J. E. Gunckel (ed.), Current topics in plant science. New York: Academic Press.

Schultes, R. E., 1969-1970, in Bull.Narcotics, 21: pt.3, 3-16; pt.4, 15-27; 22: pt.1, 25-53.
Schultes, R.E. & A. Hofmann, 1973, The botany and chemistry of hallucinogens. Springfield, Ill.: Charles C. Thomas, Publisher.
Schultes, R. E. & B. Holmstedt, 1968, in Rhodora, 70:113-160.
Schultes, R. E., B. Holmstedt & J.-E. Lindgren, 1969, in Bot.Mus.Leafl., Harvard Univ., 22:121-132.
Schultes, R. E., & R. F. Raffauf, 1960 in Bot.Mus.Leafl., Harvard Univ., 19:109-122.
Schulz, O. E., 1964, in Planta Med., 12:371-383.
Seitz, G., 1967, pp.315-338 in D. Efron (ed.), Ethnopharmacologic search for psychoactive drugs. Washington, D.C.: US Pub. Health Serv., Publ. No. 1645.
Sharma, J. D., P. C. Dandiya, R. M. Baxter & S. I. Kendal, 1961, in Nature (London), 192:1299-1300.
Shinners, L., 1965, in Taxon., 14:103-104.
Shulgin, A. T., T. Sargent & C. Naranjo, 1967, pp.202-214 in D. Efron (ed.), Ethnopharmacologic search for psychoactive drugs. Washington, D.C.: US Pub. Health Serv., Publ. No. 1645.
Silva, M. & P. Mancinelli, 1959, in Bol.Soc.Chilena Quim., 9:49-50.
Singer, R., 1958, in Mycologia, 50:239-261.
Taber, W. A. & R. A. Heacock, 1962, in Can.J.Microbiol., 8:137-143.
Taber, W. A., R. A. Heacock & M. E. Mahon, 1963, in Phytochemistry, 2: 99-101.
Taber, W. A., L. C. Vining & R. A. Heacock, 1963, in Phytochemistry, 2:65-70.
Taylor, V. E. & M. H. Malone, 1960, in J.Am.Pharm.Assoc., 49:23-27.
Taylor, W. I., 1966, Indole alkaloids. Oxford: Pergamon Press.
Truitt, E. B., Jr., 1967, in D. Efron (ed.), Ethnopharmacologic search for psychoactive drugs. Washington, D.C.: US Pub.Health Serv., Publ. No. 1645.
Tyler, V. E., Jr., 1966, in Lloydia, 29:275-292.
Tyler, V. E., Jr. & D. Groger, 1964, in J.Pharm.Sci., 53:462-463.
Usdin, E. & D. H. Efron, 1967, Psychotropic drugs and related compounds. Washington, D.C.: US Pub. Health Serv., Publ. No. 1589.
Wagner, H., 1969, Rauschgift-Drogen. Berlin: Springer Verlag.
Waser, P. G., 1967, pp.419-439 in D. Efron (ed.), Ethnopharmacologic search for psychoactive drugs. Washington, D.C.: US Pub. Health Serv., Publ. No. 1645.
Wassén, S. H., 1964, in Ethnos, 1-2:97-120.
Wassén, S. H., 1967, pp.233-289 in D. Efron (ed.), Ethnopharmacologic search for psychoactive drugs. Washington, D.C.: US Pub.Health Serv., Publ. No. 1645.
Wassén, S. H. & B. Holmstedt, 1963, in Ethnos, 1:5-45.
Wasson, R. G., 1958, in Proc.Am.Phil.Soc., 102:221-223.
Wasson, R. G., 1959, in Trans.NY Acad.Sci., Ser.II, 21:325-339.
Wasson, R. G., 1962a, in Bot.Mus.Leafl., Harvard Univ., 20:77-84.
Wasson, R. G., 1962b, in Bot.Mus.Leafl., Harvard Univ., 20:161-193.
Wasson, R. G., 1963, in Psychedel.Rev., 1:27-42.
Wasson, R. G., 1967, pp.405-414 in D. Efron (ed.), Ethnopharmacologic search for psychoactive drugs. Washington, D.C.: US Pub.Health.Serv., Publ. No. 1645.

Wasson, R. G., 1969, Soma, divine mushroom of immortality. New York: Harcourt, Brace & World.
Wasson, V. P. & R. G. Wasson, 1957, Mushrooms, Russia and history. New York: Parthenon Books.
Weil, A. T., 1965, in Econ.Bot., 19:194-217.
Weil, A. T., 1967, pp.188-201 in D. Efron (ed.), Ethnopharmacologic search for psychoactive drugs. Washington, D.C.: US Pub.Health Serv., Publ. No. 1645.
Weil, A. T., 1969a, pp.355-366 in J. E. Gunckel (ed.), Current topics in plant science. New York: Academic Press.
Weil, A. T., 1969b, in Sci.J., 5A(3): 36-42.
Wieland, T., 1968, in Science, 159: 946-952.
Willaman, J. J. & B. G. Schubert, 1961, Alkaloid-bearing plants and their contained alkaloids. Washington, D.C.: US Dept.Agr.Tech.Bull.No. 1234.

DRUG USE AND CULTURAL PATTERNING

The use of drugs as well as the kind and pattern of their use is as much a product of a particular culture as are dietary patterns or leisure time expenditure. Causally we must recognize a historical and a temporal factor, but also such influences as ethno-social contact, ecological factors, and legal prescriptions.

The chapters presented in this section reflect a widening geographical scope of drug studies, but emphasize the historical influence of setting when it is combined with cultural patterning. Thus Partridge describes the use of cannabis in the work gang in Colombia, Carter the use of alcohol among the Aymara of Bolivia, and Harding and Zinberg the way in which American cultural sanctions regulate drug abuse and legal prohibition. Du Toit views the multi-ethnic situation in South Africa as it produces different historically based patterns of drug use.

But more importantly all four of these papers address the question of ritual and whether in fact we are justified in using the concept of ritual in such a secular context. Until very recently the term "ritual" was thought to apply only to sacred contexts. In different ways each of these papers justifies its use in the secular setting of drug use.

WILLIAM L. PARTRIDGE

3

TRANSFORMATION AND REDUNDANCY IN RITUAL: A CASE FROM COLOMBIA[1]

Recent work on the role of ritual in social life seems to be divided into two distinct approaches. The first might be called the 'system transformation approach'.[2] This approach is exemplified in Rappaport's (1968) thermostat analogy which defines ritual as a regulatory device in a biotic community. From the perspective of the individual organism, Wallace's (1966) treatment of the revitalization process also exemplifies this approach. V. Turner (1969) emphasizes the transformative functions of symbols in the liminal phases of the ritual process. Chapple & Coon (1942) and Chapple (1970) approach harvest rites and other cyclical celebrations as transitional devices in the social system. The system transformation approach has as its defining characteristic the analysis of ritual as part of the process of transitional change in a larger social system, and in the examples mentioned above the instrumental, measurable ways in which rituals alter human social life are described. In this view ritual is a transformative process within a wider system of relationships; as an element in the system changes, this forces a transformation of the entire system. Ritual occurs when such changes make adjustments of the system necessary.

The second approach can be called the 'structural redundancy approach'. It is exemplified in the work of Leach (1966) who interprets rituals as condensed storehouses of a cultural tradition which through repetition and redundancy accomplish the task of blocking noisy interference and permitting communication of knowledge to occur. Similarly, Gluckman's (1954) treatment of 'rituals of rebellion' as high drama in which change is not expected and no alteration of the relations among elements results can be called a structural redundancy approach. Levi-Strauss (1963) considers rituals and their myths to be a kind of language which displays the social structure through repetitive performances. In this view rituals are restatements, stylized performances and displays of the social system as it is presently constituted. It was in this sense that Warner (1959:452) advised that we view rituals as reflections of 'a system of meaningful acts commonly shared'.

The contrast between the system transformation approach and the structural redundancy approach is real, but not absolute. As Moore & Myerhoff (1975:10) have emphasized, ritual at one and the same time marks with solemnity "change as often as it celebrates repetitions and continuities". Of particular interest here, Moore & Myerhoff emphasize situational changes within larger systematic relationships as common points of ritualization, such as a legal hearing or a birthday party. They point up the fact that even in cases of clear structural redundancy minute adjustments or transformations of relationships among elements of the parts of the system do occur.

The contrast, then, is a matter of the scope of theoretical focus. If the scope is broad and we focus upon those events which result in shifts in relationships among parts of a system, then rituals which govern these events are transformative devices (e.g. marriage rites, Dog Soldier rites before the Spring hunt, uprooting the 'rumbim'). If we narrow the scope to events which do not result in shifts in relationships among parts of a system, then rituals which govern these events are structural redundancies (e.g greeting rituals, legal hearings, Memorial Day parades).

The significant difference between the two does not seem to have to do with the function of ritual in every instance of occurrence. Rituals have various functions, depending upon the conditions in which they take place. The important distinction is made on the basis of the consequences of ritual performances of different kinds. Certain rituals result in system transformation; others do not. A birthday party is a transformative event only from the perspective of ego, not from the perspective of the vast social system of which he is part (naturally some cultures may choose a birth date as a signal for transformation, but not all birthdays are so designated). Similarly, legal hearings which solemnified the murders of black and white civil rights workers in the American south during the 1960's were clearly displays of the system as then constituted and not system transformations (of course legal hearings eventually came about which were transformative, but these particular ones and most others are not). In contrast, uprooting the 'rumbim' among the Maring is a transformative event which portends dramatic shifts in human relationships, not all of which can be predicted. To the extent that conditions vary, such as climate, fertility, etc., the consequences of this ritual is a system transformation. Similarly, the Poro and Sande bush school rites among the Kpelle of Liberia mark system transformations as neophytes are ushered into a new status and disturb the existing structure of community life. Gradually a new structure emerges which is a new system of relationships among the parts.

Yet it also seems evident that certain rituals encompass both kinds of consequences or mark both kinds of events. They stem from periodic or irregular changes in the systematic relationships among the parts of a system, but are also stylized performances of the emergent system of relationships. Marriage rites are the most obvious example of events which transform relationships among the parts yet at the same time constitute restatements of form or structural redundancies. Such rituals have a dual significance. Unlike the birthday party they are not primarily redundancies. Unlike the harvest festival they are not primarily transformations.

The contrast I am striking here is not meant to be polar. Obviously rituals celebrating events which usually have structural redundancy as their consequence may at certain, culturally defined times also entail dramatic shifts in the relationships among participants in the social system. The 'barmitzvah' and Catholic confirmation ceremonies are examples. Likewise, transformative rituals which usually have as their consequence change in systematic relationships may under certain, unpredictable conditions become mere routines which are performed out of habit but result in little change. The 'potlach' of the Northwest Coast tribes following the introduction of European trade would be an example.

The interesting issue raised by the contrast struck here is: when do rituals have the dual properties of system transformation and structural redundancy? How do they differ from other rituals? What imbues these rituals with their dual significance? Answers to these questions may permit the development of a taxonomy of rituals and obviate the need for a taxonomy of theories of ritual.

I will discuss an example of a ritual which has these dual properties. It was observed in 1972 and 1973 during fieldwork on the north coast of Colombia. It is a secular work ritual of estate workers, peasants and artisans of the lower sector of the coastal subculture (cf. Partridge, 1974, 1975). This particular ritual makes a good example for the purpose of the present analysis because it is clearly a part of the mundane, secular sphere of human affairs in this culture. Marriage rites would make an equally appropriate example, but these are often religious in nature and attention is distracted from the behavioral mechanisms at work by reference to the unknown operations of the psyche. The work of rituals of coastal Colombians will make a clearer analysis possible.

As Leach (1966:404) observed, the "orthodox convention" in anthropology is to reserve the term ritual for magical and religious behavior. The convention stems from Durkheim's (1947:38-42) distinction between the sacred and the profane. In this view everything is either sacred or profane, and the sacred is distinguished from the profane through consensual "interdictions" (Durkheim, 1947:38-42). But as Moore & Myerhoff (1975:29-30) point out, a "two-part universe" in which everything is either sacred or profane is difficult to justify ethnographically. It makes more sense to consider ritual as a mechanism or kind of behavior which operates in all spheres of social life, as a "medium in which to make statements about matters that are postulated, or unquestionable" (Moore & Myerhoff, 1975:31). Rappaport (1974) has reached this conclusion in his discussion of "sanctity". In this way Warner's (1962) Memorial Day rites, the May Day parade in Moscow, legal hearings, building dedications, and other patently secular events can be recognized as part of a class of behavior we designate ritual, together with the sacred manifestations such as shamanistic curing, divination, and other religious manifestations in the societies which first interested anthropologists. I will assume that secular rituals are the same kind of phenomena as sacred rituals, the main difference being the assertion of participants in the latter that supernatural forces are brought to bear (Wallace, 1966:107-108).

Rituals which have the dual properties of system transformation and structural redundancy share with other rituals the defining feature of a repetitive, reassertive form. As Moore & Myerhoff (1975:20) write, "ritual is a declaration of form against indeterminacy" (emphasis on 'against' in original). Ritual is an ordered statement of pattern against randomness, order against idiosyncracy as Chapple (1970), Leach (1966), Rappaport (1974) and others agree. This repetitive form is related to the function of ritual. Order in ongoing social interaction is a universal requirement of the species. Man and other animals achieve adjustment to the rhythmic cycles of the environment and the biological cycles of other animals, a process Chapple (1970:27-28) calls entrainment. This does not imply conflict-free interaction, but predictable patterns of interaction. When this state exists the animal community is said to be in homeostasis. Yet the continual changes through which individuals pass in the maturation process as well as the continual cycles of change in the external environment mean that the system is never in homeostasis. Change which calls for adjustment of interaction patterns is a constant feature of social life (Chapple, 1970:295). It is here that ritual frequently occurs. Through the repetitive, reassertive structure a ritual society is able to "reduce the variability (inherent in any change) and thus increase the probability that particular patterns will be followed" (Chapple, 1970:316, parenthetical comment mine). Entrainment is achieved, in part, through the repetition of elemental behavioral forms or interaction patterns and the contextual meanings given those forms or symbols in a ritual (Chapple, 1970:294-296). Rituals with the dual properties of system transformation and structural redundancy do not differ in this regard.

The interesting issue, then, is how certain rituals which are imbued with structural redundancy by their very form at the same time provide for system transformation. I will explore this issue in the following example of a work ritual in Colombia which involves the use of cannabis.[3]

I

The 'chagua' is a specific work ritual which occurs in a community I call Majagua on the great cattle estates, the rice estates, on peasant farms, and in the shops of brickmasons, saddle makers, carpenters and other artisans. This form of work organization has its origins in a communal work party which dates from before the conquest.

The Spanish grafted onto the native cultures of Colombia the cattle estate system and the 'encomienda', a device for recruiting labor to the haciendas. Alongside of the estate and its tribute labor stood indigenous communities which were taxed for the support of Spanish cities of the coast. Both in native communities and on the great estates the tributes and taxes were met by communal work parties; the Spanish did no more than link a traditional form of organization into the tribute system. In fact ordinances were issued which permitted Indians to conduct "their drunken feasts upon occasion of the collective planting, on the condition that there not be excess" (Patiño, 1965:293, my translation). We read the following description of the work party called the 'chagua':

> The Indians (in the jurisdiction of Santa Marta) in order to have less work in their fields have introduced a change of pace, which they call the 'chagua', in which the Indians of a town gather one day of the week, or a part of them, each with his axe or machete at the house of the Indian making the garden, and together, they clear the brush away leaving it ready to plant, while the owner of the 'chagua' is obligated to give food and drink that day, necessitating assembling much food on the part of the owner, and much maize beer on the part of his woman. This day for them is a day of rest and they treat it as a fiesta, so that it is necessary for the priest to say the mass early, and be careful that they pay attention. They return at night and if there is any drink left, form their dances until they tire. They retire then, and when one of them must sponsor a 'chagua', it is obligatory for the owner of the last one to participate since he received this benefit (Patiño, 1965:392-393, my translation).

This form of labor organization was so popular with the Spanish that hospitals, care for the aged, food for the poor, as well as the construction of buildings and work in the fields were all accomplished through the mechanism of the 'chagua' (Patiño, 1965:342-343). On the great estates the cattle barons demanded tribute work in their gardens as well as in cattle handling, and the 'chagua' was again the mechanism (Patiño, 1965:409-410).

Up until the beginning of this century this form of labor organization persisted on the north coast. Tasks such as weeding pasture (Patiño, 1965:374), transporting produce (Patiño, 1965:405, 493), subsistence horticulture (Patiño, 1965:392-393), and housebuilding (Patiño, 1965: 342-343) initially carried out by Indians were soon learned by Negro slaves who replaced the decimated native population. So thoroughly was this device learned that even among escaped slaves who formed free towns called 'palenques' the 'chagua' persisted as a device for organizing collective labor (Escalante, 1964:127).

In 1892 the United Fruit Company came to this region and the banana plantation became the predominant land use form (Kamalaprija, 1965; Economía Colombiana, 1957). Plantation monocrop agriculture had never before been typical of this region (Patiño, 1965: 503; 1969: 315), and the forms of labor organization introduced by the Company contrasted in some ways with indigenous forms. Wage labor was the mode of operation. Workers were given housing, credit at the Company store, hospital care and a wage, all of which had been customary on the great cattle estates as part of the patron-client relationship. But workers were not permitted to grow subsistence gardens on the monocrop plantations, as they had been on the cattle estates, and they became increasingly dependent upon the relatively good cash wage. While the traditional form of productive activity, the 'chagua', was displaced by modern industrial specialization among some workers,[4] the majority of workers still labored collectively in work gangs. They were called 'macheteros' and together they cleaned the land, planted banana plants, cut irrigation ditches, and cleaned and weeded the growing crop under the direction of a 'mandador' (supervisor). Thus the collective work group persisted even on monocrop plantations, but the elaborate feasting and drinking that had accompanied the 'chagua' was rescheduled. The Company schedule dictated that feasting and drinking be confined to weekends and festivals rather than at work, and the reciprocal exchange ties which had previously knit kinsmen of the community into obligatory collectivities now became less rigid and formal. Reciprocal exchanges now took place in bars, in poolrooms, in brothels and in visits to other workers' homes. Moreover, these came to increasingly encompass nonkin, since the banana boom sparked a tremendous migration into the region.[5]

In 1964 the Company completed its departure from this region. The boom days of the banana industry were over, and Majagua went through a process of economic crisis. In the last decade the land occupied by plantations has reverted to the traditional cattle estate once again, and mechanized rice agriculture has been innovated on a small percentage of the irrigated land. Peasant farms were never common in this region and are not so today. The forms of labor organization that the Company introduced did not persist unmodified. On the estates skilled employees enjoy a relationship exactly like the traditional patron-client relationship but crews of day laborers hired to ditch the fields, clear brush, and clean the irrigation canals are only paid a cash wage. Day laborers (the small number of peasants and artisans come out of the ranks of the day laborers) are hired by the job and are then dismissed. In a situation of economic depression, surplus labor, seasonal work, and low wages a migratory agricultural laboring class has become characteristic of the north coast (Bernal, 1971). Under these conditions the work gang known as the 'chagua' persists, and the exchanges of valued items which accompany work have become even more important.

It is in the work gang that cannabis use is typical. Canabis smoking diffused to Colombia only during the last 50 years (Ardila Rodriguez,

1965; Patiño, 1969:405) and became customary only in these work gangs. Life histories reveal that smoking begins at the stage of the life cycle when adolescent males adopt adult work patterns, between the ages of 12 and 22, and that initiation into cannabis smoking invariably begins in the work gangs out in the fields and pastures and not in leisure activities. Informants have between 11 and 31 years of smoking experience, and no informant reports that his father used the drug. Rather, it is members of the nonkin based work group who initiate the neophyte.

Gathering for work in the morning and during rest breaks throughout the day are the occasions for sharing cannabis. Fulltime skilled employees frequently use alcohol in a similar way, but alcohol is prohibitively expensive for day laborers. Not all workers are able to provide cannabis for themselves or others on any given day, and there exists an unstated rule that cannabis is to be shared among work group members along with the tacit assumption that others will reciprocate this favor. Over time such reciprocal exchanges grow into relationships of permanence. Conversely, the breach of the implicit exchange contract brings explicit condemnation from one's fellows in the form of gossip. Those who do not honor the subtle obligation to reciprocate cannabis soon find themselves ostracized by the label 'vivo'.

To be 'vivo' is to be considered active of mind, intelligent but unscrupulous and untrustworthy. In the context of an exchange relationship this adjective takes on a distinctly negative meaning, for it warns others against an individual. He is one who takes unfair advantage by accepting a gift, but not repaying it. The term 'vivo' is of course not used in direct address but behind the offending person's back; this is the nature of gossip and the reason for its effect.

Given the scarce job market, the rapid turnover in migratory candidates for work gangs, and the highly competitive atmosphere surrounding wage labor, an individual known as a labor broker ('el hombre que indica') has arisen. This man contracts with estate managers for a certain number of days of work of a certain number of day laborers, and it is his responsibility to have his work gang at the estate on designated days. Good labor brokers are carried by estate managers during slack times and paid a normal wage, so competition for these positions is intense and the broker is under pressure to provide reliable workers. Consequently, the labor broker and his crew distrust the man who is labelled 'vivo'. They want men who will honor their obligations, who are loyal, and the failure to purchase and share cannabis is an indication of lack of attention to the subtle social obligations that bind men together. In a situation involving nonkin based work gangs attention to such obligations is particularly important. Migration is always a possibility, for kin ties do not bind members of work gangs. Under these conditions men known to be 'vivos' are ejected from work gangs.

This does not mean that the 'vivo' man is doomed to failure. Certainly ostracism is the result, but not necessarily failure. Those who are truly intelligent, and not merely insensitive to social obligations, who prefer to risk ostracism and conserve their minimal wage may discover ways to compensate for the label and in fact turn it to their advantage. A 'vivo' man might save his earnings, forgoing secure membership on a stable work gang, invest in fighting cocks, or marketing fruits and vegetables purchased from peasants, and do quite well. Most men, however, depend upon good relations they maintain in work gangs and prefer the security of tenure on the estates. To be called 'vivo' for most is to be marked for exclusion from the form of productive activity in which the majority of adult males in the community are engaged.

It can be seen, therefore, that consumption of cannabis is linked in the ritual of reciprocal exchange to the process of recruitment of labor. Through this mechanism exclusion of unwanted members is achieved; closure of the nonkin based group is achieved. The exchange of valued items among the group of workers has always been a basic behavioral form on the north coast, but it persists today among nonkin based groups rather than kin groups.

This elemental behavioral form finds expression in other domains of culture. Two of these are housebuilding and loaning of money for purchasing alcohol for religious celebrations. In each we see the significance of the ritual in the lives of coastal workers.

Housebuilding occurs at the stage of the developmental cycle of the household which Fortes (1958) calls dispersion, when children marry and either add members to the parental house or establish a separate residence. In this community a separate residence is the ideal to be achieved, and many young men and their mates spend a few years residing with relatives before it can be attempted. During this time the young man accumulates wood, palm leaves, bricks and cement. Houses grow by accretion, as first a mud-and-stick 'bareque' is constructed with a thatched roof, then a brick and concrete block wall is erected around it, and then a brick and concrete block house replaces the 'bareque', perhaps with a zinc roof. The process may take a decade to complete, and at each stage the household head depends upon nonkin members of the work group for weekend and evening labor on his house. In addition to labor he seeks advice, the loan of tools, and materials from his coworkers. For those who contribute he will have to provide additional gifts such as food and drink, for all he provides cannabis. Of course the young man who is building a house will spend much of his time working on the slowly developing houses of his fellows.

Similarly all men of the work gang stand at one time or another as godparents to the child of a coworker. Acceptance of this honor, as in the case of marriage, death of a relative, baptism, rites of confirmation, and other religious events means sponsorship of a feast, paying fees to the Catholic priest, and assembling great quantities of alcohol. Food for the feast is not lavish, and the priest's fee is minimal. But alcohol is quite expensive for a worker (15 pesos for a 6 ounce bottle) and low wages (20 to 25 pesos a day) do not permit a man to save any money towards purchases of large amounts. If saving money for such massive expenses is not a realistic goal, maintaining a wide range of nonkin reciprocal ties of obligation is. Such ties can be had by investing only a few pesos a week in cannabis. When religious obligations culminate in a huge purchase of alcohol, a man then turns to a number of coworkers from whom he seeks to accumulate the necessary money to meet these obligations.

To be labeled a 'vivo', therefore, is a matter of consequence in these work gangs. It means to be denied assistance in achieving several of life's most valued goals. All men of the coastal lower sector seek the status of 'padre de familia', literally, father of a family. But this means much more than mere paternity; it means a man must hold a full-time job, procreate children, live in his own home, and meet his religious obligations to family and friends. To achieve this status is possible for most men only through the mechanism of winning tenure on a work gang.

In this way it can be seen that the elemental behavioral form or interaction pattern which is reflected in this ritual is the reciprocal exchange of valued items among nonkin members of the work gang. The symbols which attach to this interaction pattern will complete the description of the ritual.

II

V. Turner (1966) has advanced our understanding of ritual by postulating that symbols have three dimensions of significance: the exegetic, the operational and the positional. The interaction pattern discussed above carries meaning at each of these levels.

At the exegetic level, the variety of explanations offered by participants in the ritual in verbalizations among themselves and in response to questions, cannabis is an energizer. Cannabis is said to reduce fatigue ('quita el cansancio'), it is said to give a man energy for working hard ('fuerza'), and it is said to give a man spirit for hard work ('ánimo'). A minor theme some report is that cannabis is generally good for one's health, and some men use it in a program of health maintenance. Other themes include the relief of pain in joints and muscles as well as brewing a tea which is used to calm a crying infant. But men universally agree that the major significance of cannabis lies in its energizing properties. While cannabis smoking is relatively recent in Colombia, this theme is not. Tobacco was felt to reduce fatigue, to relieve pain and to energize workers during the colonial and republican periods of Colombian history (Patiño, 1967:295-297). It is probably the case that cannabis was substituted for tobacco at the time when it diffused to the north coast.

At the operational level of significance, the use of symbols by participants, we find that the primary significance of sharing cannabis seems to lie in the process of recruitment to the nonkin based work group and the mechanisms of gossip which regulate access. Gluckman (1963) has treated the functions of gossip most completely. Gossip functions to measure individual behavior against values held by group members, and in this manner unifies the group by excluding some and including others. More significantly in this instance, gossip enables groups to control aspiring individuals and competing cliques of which all groups are composed. This operational significance has been seen to be very much in evidence in the case of exclusion of the 'vivo' person.

At the positional level, the meaning of the symbol derived from its position in a specific cluster or 'Gestalt', we find the exchange of cannabis to be instrumentally related to some of life's most important goals. Attaining adult status which involves winning tenure on a work gang, beginning a family of procreation, building one's own house, and performing one's religious roles in family and community: these form a cluster of symbols which infuses cannabis smoking with meaning beyond the context in which it occurs. Cannabis is smoked out in the fields and pastures, but the display of cannabis smoking is of much wider importance. It is an affirmation of willingness to be transformed into a reliable male coworker, an affirmation of fitness as a trustworthy and capable adult male. Meaning is thereby transfered from the context of the work ritual to other contexts throughout the culture, and the act of smoking cannabis is made into an event of major significance.

III

Building upon this description of the interaction form and symbols which constitute the work ritual involving cannabis, the issue of the dual properties of system transformation and structural redundancy can be addressed.

The exigencies of work on the north coast of Colombia are such that a migratory agricultural laboring class has become one of the distinguishing features of the coastal subculture. Kinship relations are bilateral and a wide range of both real and fictive kin may play important roles in the life cycle of the coastal male from adolescence through old age. But kinsmen are not usually available. Under conditions of migration, low wages, labor surplus, and constant job turnover, nonkin have important roles to play. Shifts and changes in social relations are common. In the fields and pastures of the coast adolescents continually disturb existing relationships by seeking admittance to work groups, and adult men who are unsuccessful or dissatisfied swell the migratory ranks still further. The result is that the coastal male depends upon nonkin more often than he does upon kin for assistance and cooperation in achieving life's goals.

Cannabis smoking in the context of work is a device which transforms nonkin migratory workers into effective, relatively stable and cohesive work gangs. Through the development of bonds of interdependency these work gangs become relatively permanent, and the social relationships established in adolescence are extended into other areas of life as the years pass. But the structural relationships between workers as well as between work gangs and their labor brokers and estate managers which knit nonkin into stable units in one generation are necessarily differently constituted in the next generation. Son does not follow father or any other kinsman into a position on an estate, but must win that position by establishing his own reputation among nonkin. The transformative function of the cannabis smoking ritual is recurrent. Each generation new members are tested, accepted and inducted in the new system of relationships. This is of significance not only in the world of work but also in other areas of life where nonkin play important roles. Social networks which evolve out of the work gang, then, are part of a wider system of social relationships which is reconstituted every generation through the ritual device.

By its very nature the ritual also provides for structural redundancy. The repetitive form of the ritual draws together subcultural themes and values reflected in the actions of older, substantially successful adult males. These are given expression in the 'chagua', a ritual form which reflects a structure of social relationships which is historically ancient on the coast. Like generations of indigenous, Negro slave and mestizo forebearers, present day workers are separated from the powerful elite who control the land, the water the government, etc. A putative superiority of workers' values exists, in which it is assumed that co-operation, reciprocal obligations, sharing, hard work and responsibility to one's fellows are absent from the ranks of the elite but present among workers. These values are positively asserted in the 'chagua', marking both the separation of the elite and peers as well as the superiority of peer values. The evaluation of candidates for work gangs takes place through measuring conformity to these values, and by behaving in the prescribed ways the candidates affirm the structural separation, the superiority and circumscribed applicability of worker values, and the dignity of their elders and forebearers. Cannabis is linked to the positive value placed upon hard labor, and like smoking tobacco in earlier times smoking cannabis today is an act of preparation for hard work. The exchange of cannabis is linked to reciprocal labor exchange, the sharing of valued resources and cooperative alliances which have always made survival of the worker possible. The connections between cannabis smoking, reciprocal exchange of valued items, cooperation among workers, the dignity of hard work, and separation from the elite social strata make cannabis a suitable bridge between numerous tangible realities of coastal lower sector life. These realities are affirmed and restated in ritual behavior, and the positive interpretations of these reflected in the behavior of candidates who are admitted in the work group provides for structural redundancy.

An understanding of how this ritual at once provides for structural redundancy and system transformation is facilitated by V. Turner's (1975) recent work. He finds that certain symbols transcend several domains of a culture. They are readily transfered from the domestic to the economic to the religious spheres of social life. In this multiple transference property Turner finds much of the explanation for the powerful transformative properties of certain symbols (such as 'individualism' in our own society). The ritual of cannabis smoking discussed here clearly involves symbols with multiple transference properties of this kind. Their applicability to various domains of social life gives them much of their transformative power.

An important addition to V. Turner's notion of multiple transference properties of certain symbols can be made on the basis of the previous analysis. Disembodied symbols have little transformative power it would seem, unless we interpret them as irritants which stimulate anomic action. Equally important should be the corresponding interaction forms or elemental behavioral forms which connect to certain symbols. It is to the transference of certain interaction forms in 'equivalent contexts' that Chapple (1970:301) has drawn attention.

Interaction forms with multiple transference properties are those which occur in a variety of cultural domains, a variety of situations or multiple contexts. The reciprocal exchange of valued items is such an interaction form which is transfered between varying domains, situations or contexts of coastal subculture. It is transfered to the domestic sphere of housebuilding and to the developmental cycle of the domestic group through the exchange of labor, skills, tools and materials. This interaction form clearly has multiple transference properties to various domains which gives it transformative power.

In conclusion, the analysis offered here indicates that system transformation is a product of those rituals characterized by symbols and interaction forms with multiple transference properties to several domains of culture. The power to provide for system transformation seems to lie in the connections and linkages which participants recognize in other spheres of experience. Cannabis and the reciprocal exchanges which surround it are equally encompassed by this process of transference between cultural domains. This gives to the ritual the dual properties of system transformation and structural redundancy.

NOTES

1. This report is based on fieldwork sponsored by NIMH grants 1F01MH54512-01 CUAN and 3F01DA54512-01S1 CUAN. An earlier version of this paper was read in the symposium 'Secular Ritual and Altered States of Consciousness'

at the annual meetings of the Society for Applied Anthropology in 1975, Amsterdam, Holland. This version resulted from a revision stimulated by reading 'Secular ritual: forms and meanings' by Sally Falk Moore and Barbara G. Myerhoff, a volume based on Burg Wartenstein Symposium Number 64, which took place in Austria in August, 1975. I wish to thank Sally Falk Moore, Barbara G. Myerhoff, Solon T. Kimball, Eliot D. Chapple and Brian M. du Toit for their comments on the revised version.

2. Transformation plays a role in both the persistence and modification of cultural systems. In one sense, transformation refers to the process of transition within a single temporal-spatial context and set of conditions, as in the transition from boy to man in a rite of passage. T. Turner (1975) considers the transformation of variables through the interplay between structural levels from this perspective. In another sense, transformation refers to developmental change which spans more than a single temporal-spatial context and set of conditions representing change in the contexts and conditions themselves. Kimball & Partridge (forthcoming) have considered such a case in the transformation of coastal Colombian culture following the coming of the United Fruit Company. In this paper I use the concept of transformation in the first sense, that of transition between states of a system within a single temporal-spatial context and set of conditions.
3. I have not used the scientific name because at present the question of one or several species is open. See Schultes' (Schultes, et al., 1975) discussion of the problem.
4. 'Cortadores' cut banana stalks, 'portadores' hauled and loaded them on gondolas, 'tanquepes' washed, sealed and bagged them, 'canaleros' controlled the flow of water to the plants.
5. Majagua grew from a score of families in 1876 (Vergara y Velasca, 1901) to a municipality of 15,861 in 1938 (Contraloría de la Republica, 1941).

BIBLIOGRAPHY

Ardila Rodriguez, Francisco, 1965, Aspectos medico legales y medico sociales de la marihuana. Tesis Doctoral Universidad de Madrid, Facultad de Medicina.

Bernal, Segundo, 1971, Algunos aspectos sociologicos de la migración en Colombia. In Ramiro Cardona Gutierrez (ed.), Las migraciones internas, 51-101. Bogotá: Editorial Andes.

Chapple, Eliot D. & Carleton S. Coon, 1942, Principles of anthropology. New York: Holt.

Chapple, Eliot D., 1970, Culture and biological man: explorations in behavioral anthropology. New York: Holt, Rinehart & Winston.

Contraloría General de la Republica, 1941, Censo general de población, 5 de Julio de 1.938, Tomo IX. Bogotá: Imprenta Nacional.

Durkheim, Emile, 1947, The elementary forms of religious life: a study in religious sociology. Translated from the French by J. W. Swain. Glencoe, Illinois: The Free Press (original 1915).

Economía Colombiana, 1957, Monografía económica del Magdalena y de la Industria del Banano, 12(35):275-629. Bogotá.

Escalante, Aguiles, 1964, El Negro en Colombia. Monografías Sociológicas Numero 18. Bogotá: Universidad Nacional.
Fortes, Meyer, 1958, Introduction. In Jack Goody (ed.), The developmental cycle in domestic groups. Cambridge Papers in Social Anthropology Number 1.
Gluckman, Max, 1954, Rituals of rebellion in South-east Africa. Manchester: Manchester University Press.
Gluckman, Max, 1963, Gossip and scandal. Current Anthropology, 4(3):307-316.
Kamalaprija, V., 1965, Estudio descriptivo de la estructura del mercado del banano Colombiano para la exportación. Bogotá: Instituto Latino-americano de Mercadeo Agricola.
Kimball, Solon T. & William L. Partridge, The craft of community study (forthcoming).
Leach, E. R., 1966, Ritualization in man: ritualization in man in relation to conceptual and social development. Philosophical Transactions of the Royal Society of London, Series B, Biological Sciences, Sir Julian Huxley (ed.), 251:403-408.
Levi-Strauss, Claude, 1963, Structural anthropology. New York: Doubleday.
Moore, Sally Falk & Barbara G. Myerhoff, 1975, Secular ritual: forms and meanings. In Sally Falk Moore & Barbara G. Myerhoff (eds.), Secular ritual (forthcoming).
Partridge, William L., 1974, Exchange relationships in a community on the North Coast of Colombia with special reference to cannabis. Ph.D. Dissertation, University of Florida.
Partridge, William L., 1975, Cannabis and cultural groups in a Colombian Municipio. In Vera Rubin (ed.), Culture and cannabis, pp. 147-172. The Hague, Mouton.
Patiño, Victor Manuel, 1965, Historia de la actividad agropecuaria en America Equinoccial. Cali: Imprenta Departamental.
Patiño, Victor Manuel, 1967, Plantas cultivadas y animales domesticos en America Equinoccial: plantas miscelaneas, Tomo III. Cali: Imprenta Departamental.
Patiño, Victor Manuel, 1969, Plantas cultivadas y animales domesticos en America Equinoccial: plantas introducidas, Tomo IV. Cali: Imprenta Departamental.
Rappaport, Roy A., 1968, Pigs for the ancestors: ritual in the ecology of a New Guinea people. New Haven: Yale University Press.
Rappaport, Roy A., 1974, Obvious aspects of ritual. Cambridge Anthropology, 2(1):3-69.
Schultes, Richard Evans, William M. Klein, Timothy Plowman & Tom E. Lockwood, 1975, Cannabis: an example of taxonomic neglect. In Vera Rubin (ed.), Culture and cannabis, pp. 21-38. The Hague: Mouton.
Turner, Terence S., 1975, Transformation, hierarchy, and transcendence: a reformulation of Van Gennep's model of the Structure of Rites de Passage. In Sally Falk Moore & Barbara G. Myerhoff (eds.), Secular ritual (forthcoming).
Turner, Victor W., 1966, The syntax of symbolism in an African religion. Philosophical Transactions of the Royal Society of London, Series B, Biological Sciences, Sir Julian Huxley (ed.), 251:295-303.

Turner, Victor W., 1969, The ritual process: structure and anti-structure. Chicago: Aldine.
Turner, Victor W., 1975, Ritual as communication and potency: an Ndembu case study. In Carole E. Hill (ed.), Symbols and society. Proceedings of the Southern Anthropological Society Number 9.
Vergara y Velasca, F. J., 1901, Nueva geografía de Colombia, Tomo I. Bogotá: Republica de Colombia.
Wallace, Anthony F. C., 1966, Religion: an anthropological view. New York: Random House.
Warner, W. Lloyd, 1959, Living and the dead: a study of the symbolic life of Americans. New Haven: Yale University Press.
Warner, W. Lloyd, 1962, American life: dream and reality. Chicago: University of Chicago Press.

BRIAN M. DU TOIT

4

ETHNICITY AND PATTERNING IN SOUTH AFRICAN DRUG USE

The use of stimulants, hallucinogens and other forms of mind altering substances are of the same order, though their results are of a different nature, than foods and drinks used by people the world over. For this reason any study of such mind altering substances must examine their use in social settings and within the cultural context. Only then will it be possible to fathom the complexity of factors which influence not only the reasons for the use of these substances but also the ways they are used. In short, uses of substances which produce altered states of consciousness must be seen as the product of a unique cultural, historical and ecological setting. Once we fully understand the use of such substances in clearly defined socio-cultural settings, it will be possible to compare cross-culturally or across other kinds of boundaries; ecological, geographical or psychological.

This paper is concerned with the use of various kinds of mind altering substances employed in South Africa. Due to the ethnic complexity which has been magnified by that country's legal system, it will be necessary to outline briefly the actual field research and the ethnic situation. Following that we can turn our attention to the patterns of use of these various substances.

THE SOCIO-CULTURAL SETTING

Following various negotiations with governmental authorities and police, the research on which this paper is based, was initiated in August, 1972.[1] The field study period concluded in June 1974. The focus of our study was the east coast of South Africa, specifically the province of Natal, where the population, in general, shows the same variation found in other parts of that country. Natal, however, is unique in certain respects. The degree to which these factors influence the use of mind altering substances will become clear in our subsequent discussion.

We concentrated the research on two samples of Zulu speaking Africans: one living in the urban patchwork and port city of Durban, the other being rural and scattered over small towns, white owned farms and plantations, and African reserve areas. About halfway through the field study it was decided to include the other ethnic groups living in the area. We employed exactly the same interview schedule and research procedure for all subjects with the exception that interviews were always administered by research assistants of the same ethnic group as the respondents. The project files contain detailed information on more than 1500 Africans (mostly Zulu speakers), 124 Coloureds, 119 Whites, and 55 Indians.[2] The criterion for including these persons was their use of cannabis.[3]

In considering the cultural background of a person in any one of these legally defined groups, it is necessary to outline a few salient historical factors.

The African population, living as pastoralists and cattle herders had drifted into the region four centuries ago. As far as we now can ascertain they had already been using cannabis,[4] tobacco for snuff, and some form of fermented drink much as they still use the traditional beer. Other than herbal potions used as medicine or for ritual purification, the Africans did not use any other stimulant or product to affect their level of consciousness.[5] Tobacco originally had been used as snuff but increasingly people were smoking it or using the smoke medicinally or ritually. In time, too, as people increased contact across social barriers, Africans started using brewery produced beer, wines and spirits. Terminologically though, Africans still differentiate between traditionally brewed beer, which the Zulu call 'tshwala' and 'beer' which is brewed and bottled commercially. Cannabis is used by members of the African population group in two ways: it is either smoked in some variation of the old hubble-bubble-pipe (see du Toit, 1975), or nowadays is smoked in a rolled cigarette form by younger users. In a few cases it is smoked like pipe tobacco in a pipe. Cannabis is hardly ever eaten, brewed, or used in any other form.

The second major population group in the region under discussion are the East Indians. The first Indian laborers arrived in Durban on November 16, 1860 aboard the 'Truro' and during the years that followed more than 6,000 persons were imported to work on the Natal sugar plantations (Pachai, 1971:6). For more than a century they have remained relatively homogeneous groups united by bonds of language and religion. These people have also been characterized by the strong sense of kinship which binds members together and by the moulding and guiding influence of the family. Due basically to the uniting role of religion and the family, the Indian community has been singular in the low incidence of divorce, secularism, problems related to youth, alcoholism and drug use. This does not imply that these aspects have been totally absent but that they have been exceptionally infrequent or culturally patterned within the range of acceptability. Major changes have occurred over the past fifteen years.

The use of alcohol for instance was curbed both by religion and law. Both the Hindu and Islamic religions forbid the use of alcohol (Balkisson, 1973: 6-7). Indians outside Natal were also prevented from purchasing alcoholic beverages by Section 95 of Act No. 30 of 1928. This restriction was lifted in 1963. But in spite of secularization and the removal of legal restrictions, Indian women rarely smoke or use alcohol. Two categories have to be excluded from this generalization, says Meer (1969: 71). Elderly women among the peasant and working class families have appropriated, and are granted, the masculine right to smoke and drink. The same permissiveness pertains to westernized women who have attained high professional qualifications. It is quite conceivable that female university students who are aspiring to this latter category, would also be persons who use alcohol and possibly cannabis.

The use of cannabis, however, has been present since their ancestors came to Natal. Law No. 2 of 1870, called the 'Coolie Law Consolidation', among other things prohibited "the smoking, use, or possession by, and the sale, barter, or gift to, any Coolies whatsoever, of any portion of the hemp plant (cannabis sativa); and authorizing the destruction thereof if found in such use or possession; and imposing penalties upon Coolies using, cultivating, or possessing such plant for the purpose of smoking the same".

Some years later, between 1885-87, the Indian Immigrants Commission or so-called Wragg-Commission (named after Walter Thomas Wragg, Supreme Court Judge) looked into the conditions of the Indian community. Their report in 1887 states:

> Some Immigrants smoke it almost in its green state, at the time (May) when it is gathered and when its active principle (cannabin resin) and volatile oil are most powerful and particularly obnoxious to the consumer. Some smoke the dried leaves only. Other Indians consume a mixture compounded of tobacco, opium, hemp, and brown sugar ... (1887: 6)

The commission points at the fine tobacco grown in Natal and recommends that the Indians be urged to smoke that instead of cannabis and opium. But they also report that "in some parts of the Colony, white traders purchase green hemp leaves from Kaffir growers and retail them, in a dried state, to any customer who applies for them" (1887: 8). This brings us then to two further subjects: the White ethnic group and the network of distribution, the latter of which we will refer to later.

The White population of Natal, unlike the rest of South Africa, was for many years English speaking both in its urban and largely in its rural population. Whites had a long tradition of using alcoholic beverages, particularly liquor and beer. There is in fact one variety of beer which states on its label that it is 'the beer that Natal made famous'. While they had close social contact with Indians and Africans alike it was extremely rare for White youths to smoke cannabis. The stigma of a 'dagga smoker' was something most Whites would avoid, particularly in such a color-caste oriented country. When White youths experimented it was with tobacco and alcohol. Once again the past fifteen years have seen some major attitudinal and behavioral changes. Due in part to a more enlightened view on the ethnic minority groups and in part on the influence of the world wide 'awakening' among youth, many young people turned to cannabis as an easily accessible and inexpensive alternative in their attempts to dissociate themselves from their situations. A great deal of international communication and mobility also brought Whites into contact with other ways of 'turning on' or 'dropping out'. But the most critical difference, in my thinking, is that the Whites had a cultural tradition which could readily accommodate and explain the use of alcohol and possibly some other drugs, but they lacked the historical acceptance and frame of reference for cannabis. The African today explains that the cannabis leaf is the 'ugwayi abadala' (the smoke of the ancestors), while the Indian has centuries of family and cultural traditions which account its use — whether this was fully sanctioned or not. In fact, the Indian Hemp Drugs Commission of 1893-94 found that in certain parts of India the cannabis plant was an object of worship, or at least of veneration (Kalant, 1972:82). Of the four major ethnic classifications in South Africa, this sort of historical tradition is lacking among both Whites and Coloureds.

The Coloureds of Natal actually constitute only a small percentage of the Coloureds in South Africa, the vast majority of this group living in the Cape Province. These people are truly a marginal people (Dickie-Clark, 1966). Ethnically and politically they fall between Whites and Africans, their principal parental groups, yet they have no language which is their own and their cultural tradition is basically that of the Whites. The largest percentage of Coloureds speak Afrikaans and share a heritage with this group, yet are not recognized or accepted by them. It is then not strange to expect that the Coloureds in their pattern of alcohol and drug use will fall somewhere in between. The Coloureds who live in the Cape, the wine region of South Africa, speak mostly Afrikaans. Thus the Afrikaans-speaking Coloureds drink more wine and are generally heavier drinkers than the English-speaking Coloureds (Venter, 1965:87).

It was among members of these four ethnic groups that research was conducted. But the members of each group must be viewed first of all in terms of their own cultural and historical background and only then in comparison with other residents of South Africa. Too often in the past White norms have been summarily applied to all four ethnic groups without taking these cultural-historical factors into consideration.

2. ALTERED CONSCIOUSNESS AND RITUAL

This paper will combine two major concepts, namely altered states of consciousness and ritual. By altered states of consciousness I refer to a psychobiological state in which the subject, due to auto-suggestion, spirit possession or the use of some preparation (for ease of reference here termed a drug) influences his perception of time and space. Obviously each of these concepts will be subject to cultural interpretation and many such states will permit direct observation and recording while others might require institutional observation. The fact that external conditions e.g. malnutrition or social expectations might influence the susceptibility of the individual is clear. Clear too is the fact that altered states of consciousness will vary both individually and culturally. However, we can expect to find patterning both in methods recognized and substances employed as we move from one community or cultural group to another. For this paper the definition given above will apply to the use of cannabis, alcohol, and a number of other drugs to be discussed. It will also apply to different cultural communities e.g. Coloured, White, and Indian but also rural African and urban African.

2.1 Dissociational states

It has already been suggested that historical patterns of cannabis use are reflected in current justifications for smoking. But there are other reasons for viewing the members of a particular ethnic category without reference to other ethnic groups.

Writing some years ago about 'African' psychiatry, Mackay observed that as workers and researchers in Africa we do not have

> a Normal for our basis, because we have never taken the trouble to study African normality ... We have so far judged our cases on their departure from a European Normal, if we have judged them at all. Or else we have judged them on their departure from a Normal which we do not know (1948:1-10).

This same criticism underlay the decision of an urban academician to take a position in a rural area in order to undertake an ethnological and psychiatric investigation of a traditional ethnic group in the eastern Cape (Laubscher, 1937). In both these cases the writers wanted to understand Normal and Abnormal in terms of the traditional background against which individual members acted. This would be the logical context for understanding psychological problems and treating them. The same approach has been emphasized and put into practice in Nigeria where a project of community psychiatry has been developed, for, says the director T. A. Lambo, "concepts of health within the framework of African culture are far more social than biological" (1964:446).

Factors such as cultural values, the importance of the family, and the principles of social structure must guide any study concerning not only general health and mental health, but also regarding dissociational states and the use of various elements to induce these. The important role of the 'sorcerer', positive in many cases (du Toit, 1971), must be kept in mind. Says Carothers:

> In general it seems that the rather clear distinction that exists in Europeans between the 'conscious' and 'unconscious' elements of mind does not exist in rural Africans. The 'censor's' place is taken by the sorcerer, and 'splits' are vertical, not horizontal. Emotion easily dominates the entire mind; and, when it does, the grip on the world of things is loosened, and frank confusion takes the place of misinterpretation (1953:161).

But this approach also applies to the other ethnic groups in South Africa. Whether one is dealing with aspects of social psychiatry (Opler, 1956: 5-24) or with hallucinatory experiences (Wallace, 1959; Bourguignon, 1970, 1973), social and cultural factors influencing the subjects must be the first concern. Of concern, too, are the ways in which members of the particular society interpret and respond to hallucination. A person's own interpretation will be a function of cultural expectations, social mores, legal implications, and valued themes present in his community.

> The function should determine in part his emotional experience both during and after the event, and possibly (by cultural suggestion) its perceived content as well (Wallace, 1959:67).

Stated even more specifically, there are at least three factors to be kept in mind over and above the pharmacological properties of the substance being used: the context in which the drug is taken and expectations which accompany this setting; the context and result of previous experiences; the type of society into which the drug experience is integrated and the success of this integration (Bourguignon, 1970:189).

This cautionary note applies to the researcher and writer but is also a plea to administrative and governing bodies. There is a need for uniformity in discussion and treatment of varying situations, but this can come only after a thorough understanding of each situation. An ethnically diverse country must recognize the constitution and background of each diverse component. For this reason it is essential to view the White drug user within the context of White society with the religious precepts, traditions, mores, values and other prototypes which pertain. But as Whites are evaluated within and against their historically derived society, so too the other ethnic groups must be seen against their historical traditions and values. Not only will this lead to greater understanding of the use of mind altering substances but it will facilitate immensely just administration and rehabilitation.

2.2 Rituals and rites of transition

Returning to the South African scene, it should be kept in mind that the use of alcohol is legal for all ethnic groups and that alcohol is in fact readily available. Until a little over a decade ago Africans could not freely buy alcohol and according to Act No. 30 of 1928 the alcoholic content of the traditionally brewed beer was restricted to 2 per cent by volume. In time the Government appointed a commission of inquiry and their positive report resulted in Act No. 72 of 1961 which permitted Africans to freely purchase alcoholic beverages.

The same is, however, not true for other mind altering substances. Cannabis had been used by mature and aged Africans according to culturally standardized patterns (vide du Toit, 1976) for centuries. Indians had used it along with opium though the use of neither substance was legal. Drug use among Coloureds and Whites did not affect a significant percentage of these populations. Under provisions spelled out in Articles 61-70 of the Medical, Dental, and Pharmacy Act, No. 13 of 1928 it became illegal for any person to grow, sell or supply in any form products denoted as "habit forming drugs". This included cannabis. In May 1971 the 'Abuse of Dependence-Producing Substances and Rehabilitation Centres Act' became law. In the Schedule Part I, a list of 'Prohibited dependence-producing drugs' appears which includes: Cannabis, Heroin, Lysergide, Mescaline, Prepared Opium and others. This law allows for a five year jail term on first conviction and a fifteen year term for subsequent convictions. It also aims at halting the trafficking in drugs.

The person in South Africa who, for whatever reason, desires to achieve a dissociational state or to have a mind altering experience[6] has two avenues open: the legal use of alcohol or the illegal use of a variety of drugs. Research has shown that a large number of subjects are willing to select the illegal drugs simply because they dislike the effects or results of alcohol. Because of the potential danger of informers, police raids, or normal arrests the person who takes the drug route subjects himself to severe strain. We must recognize that the first use of a highly illegal drug (as they all are) represents for a person a change of status, a change from a status of relative immunity from the law to a status where arrest (and prosecution) can follow at any moment.

We should recognize here what Arnold van Gennep called 'rites de passage'. As Gluckman has pointed out,

> van Gennep's thesis was on the study of the mechanism of rituals, rather than on the role which whole ceremonies and specific rites play in the ordering and re-ordering of social relations (1962:4).

At this stage of our discussion I also am basically interested in the mechanism of these rituals of transition and their implications. A person who smokes a single puff of cannabis is legally as guilty as a person found using heroin or LSD.[7] The decision then to use even cannabis represents a change in the attitude and future associations of the user. Repeatedly subjects would explain that their previous friends don't react against their use of drugs because they now 'hang out with a different lot'. Or in cases where persons had changed their drug habit the statement most frequently given is that the subject is no longer associates with a certain group of persons. Such a person then had in fact participated in a ritual, the component elements of which are group specific, and had set himself apart from previous associates. Because the ritual elements are different for various ethnic groups one usually finds ethnically homogeneous sets of users – persons who have much the same cultural background and expectations. The most important aspect of this ritual is that it constitutes an attitudinal change and an interactional change. This does not imply that the user cannot discontinue his use or return to a previous interaction set, but it can only be achieved by discontinuation of use and disassociation with the interaction set. But van Gennep dealt with the mechanism of rituals of transition and it is to these rituals I must now turn.

The distinction between sacred and secular rituals has been successfully muddied in recent years. Those who prefer to speak only of sacred rituals, suggesting that the latter concept is defined only by reference to the former, like to quote Emile Durkheim. After all he was the person who gave us the useful dichotomy between sacred and profane and, it is suggested, rituals can pertain only to the former. In the first chapter of 'The elementary forms of the religious life', Durkheim discusses sacred things. He then continues: "A rite can have this character; in fact, the rite does not exist which does not have it to a certain degree" (1954:37). We are left to ponder what the author had in mind when he suggested that a rite can have a sacred character. I would suggest that Durkheim was defining ritual as something set aside not because it occurred only with reference to "personal beings which are called gods and spirits", but because it pertained to a specified place and the condition of the actors. For this reason the author could later in this same book (while discussing delirium and religious exaltation) state:

> The ritual use of intoxicating liquors is to be explained in the same way. Of course this does not mean that an ardent religious faith is necessarily the fruit of the drunkenness and mental derangement which accompany it; but as experience soon informed people of the similarities between the mentality of a delirious person and that of a seer, they sought to open a way to the second by artificially exciting the first (1954:226).

Various mind altering substances may be used in this context or may be used by persons for completely different and individualistic reasons.

The condition is that the performance of the act, which we define here as a ritual, be limited to persons who by its performance change their status and that certain regularities or patterning apply to the act. Thus, when persons perform standardized acts in first using cannabis, they change from non-drug users to users, from persons who are immune from arrest under the drug law to potential convicts, from observers to participants.

Approaching this problem from a different angle we might accept that the old view that ritual necessarily had a religious or sacred context or referrent has been replaced by a modern interpretation. Ritual then refers here to "a category of standardized behavior (custom) in which the relationship between the means and the end is not 'intrinsic' i.e. is either irrational or non-rational" (Goody, 1961:159). This then allows for a study of "ritualization of women's life-crises" (La Fontaine, 1972) to appear in the same volume as a study of "verbal and bodily rituals of greeting and parting" (Firth, 1972). It also differs from the position taken by Turner who defines a ritual as "a stereotyped sequence of activities involving gestures, words, and objects, performed in a sequestered place, and designed to influence preternatural entities or forces on behalf of the actors' goals and interests" (1973:1). The ritual of cannabis use entails activities, gestures and objects, and may also involve words but it is not aimed at any agent, entity or force. The satisfaction of the act is in its completion; it is an act which serves to give unity, identity, and transition to the participants. Its value is in its performance.

Accepting Goody's definition of ritual, which we can apply here to the secular realm of cannabis use, we are in fact faced with two levels or three contexts of application: The ritual of actual use (which may have a traditional or a modern context) and the ritual of first use.

1. The traditional use of cannabis — Most of the African societies in southern Africa had established ritual uses of cannabis. The Zulu, as a case in point, restricted its use to mature and usually older men. In these cases two or more men would meet after a day's activities and prepare a hubble-bubble, usually made with a clay bowl and a water container of ox or antelope horn. The saliva which formed in the mouth was passed through "a hollow stem of tambootie grass and so made to trace a labyrinth ('tshuma sogexe') on a smooth floor" (Samuelson, n.d.:81). This could develop into a competition as one person prepared a labyrinth and the other attempted to find his way through it, similarly using his grass tube and saliva. It was a ritualized way of relaxing, of looking back over the day, of reminiscing about the past.

Another ritual use in traditional society was prior to the departure of warring parties. The Zulu had one of the most feared regimental systems. Prior to an attack the young warriors smoked cannabis and were then fearless in their pursuit of victory.[8]

2. Everyday use of cannabis — In both the traditional and the modern use, there is a particular stylized ritual. This obviously will differ from one ethnic group to the other, but among Africans it involves the handling, grinding, preparation, mixing with tobacco and smoking of the cannabis. Again this will differ only depending on whether the substance is smoked in a water pipe, in a regular pipe, or in a rolled cigarette form. Not only do we find a standard ritual in grinding the cannabis in the palm of the left hand with the right thumb, followed by the removal of all stalks and seeds, but there are also common ratios of mixing tobacco to cannabis. Certain groups in the African ethnic group also have prescriptions of which kind of tobacco should be used for the mixture. While Whites use cannabis in the pure form, Africans in nearly all cases explain that smoking pure cannabis will result in 'imphehlwa' — a deepeating ulcer.

3. The ritual of first use — As suggested earlier in this discussion, the first use of a highly illegal mind altering substance will affect the life of the user. Even if he is a loner, using the substance only when alone, he still has to acquire it, and in time must establish a dependable network of supply.

This ritual of first use really is a psychological breakthrough, a willingness, a commitment. "Almost every human action that takes place in culturally defined surroundings is divisible in this way; it has a technical aspect which does something and an aesthetic, communicative aspect which says something" (Leach, 1968: 523). The use of this substance is also going to influence the user's interactional network. Thus it must be seen as a 'rite de passage'.

Emphasizing the social rather than the mechanical aspects of rites of passage, Chapple & Coon (1942: 484-528) show how each of the three stages consists of a change in interaction rates. In other words, the frequency and intensity of a person's interaction with certain other persons will change as he progresses from one stage to the other. I would suggest that as a subject moves from a stage of non-smoker, or non-user, to the stage where a particular substance is used, he will discontinue earlier associations and establish a new or changed social network consisting of suppliers and fellow users.

A White cannabis user may go so far as to have the right hand, just above the thumb, tattooed with five small dots and one dot below his nose. He may also wear the cult ring, depicting a rocket on a trip to a cloud, on the index finger of the right hand. An African user may allow the nail of the third or fourth finger on the right hand to grow over one quarter inch in length as this is useful in picking out the stalks and seeds, but also marks him as a user.

The user may be less obvious about his new status, and this might involve only the learning of the cult-jargon terms and references by which users and non-users can be differentiated. Here again there are clear patterns distinguishable among the different ethnic groups. And while supply networks and the groups that use drugs may cross ethnic lines, it is more common to find ethnically homogeneous groups of users.

3. DESCRIPTION AND ANALYSIS OF A SAMPLE OF CANNABIS USERS

This section will deal specifically with a quantitative treatment of a sample of cannabis users as they relate to the use of various kinds of mind altering substances.[9] The sample includes the total number of subjects for the Coloured, Indian and White groups, but only 196 of the Africans drawn by numerical random sampling from the total of about 1500 subjects. The sample includes males and females, but it should be pointed out that relatively few African and Indian and few Natal Coloured women use substances to alter their states of consciousness.

The criterion for inclusion in this sample was the fact that the person used cannabis. He may have used other kinds of stimulants or consciousness altering substances as well, but all persons referred to here were selected because they were cannabis users. In Table 1 this sample has been described by ethnic group membership, by sex, and by age. In establishing the age categories I saw seventeen as the normal school-leaving age after which persons could either go to college or to work. They are in either case much more independent after leaving high school. The second category might include the most active age group, persons during their college or university years and young persons who have recently started to earn their own income thus becoming economically idependent, and usually persons who are single. The third category might include young married persons and persons who are settling down in jobs. They are gaining increasingly more responsibilities and obligations, but are a group in transition. The last age category includes all the 'over thirties'. In addition to differentiating these persons frequently thought of as 'older' or 'square' or 'adult' we are, of course, dealing with persons who are removed by at least fifteen years from that critical age when persons frequently start experimenting with altering their states of consciousness. The fifteen intervening years for each ethnic category represents critical changes which have come about and which represent legal as well as social-cultural factors.

The initial breakdown also contrasts the 17.2 per cent of our sample who have used other substances with the 78.1 per cent who use only cannabis and have never tried anything else. In the following analysis each of these statistical categories is then analyzed by the factor of age. Without reference to ethnic group membership, we see that for the below-seventeen-years-old category 42.6 per cent have tried other drugs in contrast with only 11.0 per cent for the over thirty category. Persons

Table 1: Statistical description of an inter-ethnic sample population of cannabis users (N = 494)

A. Ethnic group

African	196	39.7 per cent
Coloured	124	25.1 per cent
Indian	55	11.1 per cent
White	119	24.1 per cent

B. Sex

Male	432	87.5 per cent
Female	59	11.9 per cent
Nr	3	.6 per cent

C. Ethnic groups distributed by age category

	African	Coloured	Indian	White	Total
Below 17	14	29	2	9	54 (10.9 per cent)
18-23	56	55	40	75	226 (45.7 per cent)
24-29	79	24	4	28	135 (27.4 per cent)
Over 30	47	16	3	7	73 (14.8 per cent)
Nr	0	0	6	0	6 (1.2 per cent)

D. Tried drugs other than cannabis

Have (ever) tried other drugs	85	17.2 per cent
Have not (ever) tried other drugs	386	78.1 per cent
Nr	23	4.7 per cent

E. Tried other drugs by age group

Below 17:	Yes	23	42.6 per cent
	No	28	51.9 per cent
	Nr	3	5.6 per cent
18-23:	Yes	43	19.0 per cent
	No	173	76.5 per cent
	Nr	10	4.5 per cent
24-29:	Yes	10	7.4 per cent
	No	119	88.1 per cent
	Nr	6	4.5 per cent
Over 30:	Yes	8	11.0 per cent
	No	61	83.5 per cent
	Nr	4	5.5 per cent
Nr:	Yes	1	16.7 per cent
	No	5	83.3 per cent
	Nr	0	—

below age twenty-four contrast with the older categories in rates of usage of other drugs by 23.6 per cent to 8.6 per cent.

The logical question which follows is the nature of these 'other drugs' which are employed in producing dissociational states. Table 2 presents this information for each of the four ethnic groups, even though the substances are simply listed here by the names which our research subjects used. The greatest amount of experimentation is with LSD where we find 33 persons or 6.7 per cent of our total sample have used it. Of these, 15 persons, or 45.5 per cent of those who tried LSD, have discontinued its use while 18 or 54.4 per cent are still using this hallucinogen. Mandrax is the next in the frequency of its use as 27 persons, or 5.5 per cent of our total sample have used it. At the time of the research 6 persons, or 22.2 per cent of those who tried mandrax had

Table 2: Statistical description of a sample of cannabis users by ethnic group who have tried or use other drugs[10]

	African			Coloured			Indian			White			Total		
	Tried	Used	Total	Tried	Used	Total	Tried	Used	Total	Tried	Used	Total	Tried	Used	Total
LSD				12	7	19				3	11	14	15	18	33
Mandrax				5	6	11		3	3	1	12	13	6	21	27
Benzene				3	21	24				1		1	4	21	25
Petrol				2	17	19				1		1	3	17	20
Opium					7	7		1	1	1	3	4	1	11	12
Amphetamines				1	4	5					5	5	1	9	10
Purple heart				2	4	6				1		1	3	4	7
Glue		1	1	1	3	4		1	1				1	5	6
Dexedrine				1	2	3				1	1	2	2	3	5
Hashish					3	3					1	1		4	4
Heroine				1		1				1	1	2	2	1	3
Black bomb				2	1	3							2	1	3
Tranquilizers											3	3		3	3
Morphine								1	1		1	1		2	2
Diet pills											2	2		2	2
Serropax										1	1	2	1	1	2
Disperin & coke					1	1				1		1	1	1	2
Barbiturates											1	1		1	1
Sleeping tabs											1	1		1	1
Orange heart										1		1	1		1
Bensodrex										1		1	1		1
Anacronol										1		1	1		1
Artane										1		1	1		1
Obex										1		1	1		1
Pendorex										1		1	1		1
Librium								1	1				1		1
Total	0	1	1	30	76	106	0	7	7	18	43	61	48	127	175

discontinued while 21 persons, or 77.8 per cent still used it regularly. Together Benzene and petrol, both of which are inhaled or smelled, constitute a very large percentage of consciousness altering substances. Forty-five persons, or 9.1 per cent of our sample have used these 'whiffers' and only 15.6 per cent of them have discontinued their use, leaving 84.4 per cent who still use these substances.

Opium is used quite commonly as 12 persons, or 2.4 per cent of our total sample have smoked opium. We find here, however, a very high rate of continued use as only one of these persons has discontinued its use leaving 91.7 per cent as regular users. This same high rate of continued use is found among the ten persons or 2.0 per cent who have used amphetamines. Once again only one person has discontinued or only tried it experimentally leaving a 90 per cent continued use.

Purple hearts, glue sniffing, and dexedrine experimenters each represent between 1.5 per cent and 1.0 per cent of our total sample. Rates of continued use vary from 57.1 per cent for purple hearts to 80 per cent for glue.

While LSD has had the greatest number of triers and users these are almost evenly distributed between Whites and Coloureds with complete absence of use among Africans and Indians in this sample. Mandrax has the widest ethnic distribution of use among Whites, Coloureds and Indians while being completely absent among Africans. It also has a high rate of continued use.

Benzene use is almost exclusively found among the Coloured subjects of our sample. The one White who tried it experimentally discontinued its use. The same is true for petrol as once again a single White experimented with this whiffer. In both cases Africans and Indians are absent from these totals. The pattern of opium use shows one White discontinuing while all the Coloured and Indian users still use it. As has been the case so often in our discussion above, amphetamines are restricted to Whites and Coloureds, most of whom still use it.

The positive correlation between ethnic group membership and the tendency to use other drugs is clear as seen in Table 3. This relationship is confirmed using Chi-square at the 0.01 confidence level. Once again the White and Coloured ethnic groups are most predictable. Reporting on a questionnaire survey of 1327 persons of Indian ancestry, Millar (1972:8) found that of the total number of respondents "19.2 per

Table 3: Tried other drugs by ethnic group

Tried other drugs	African	Coloured	Indian	White	Total
Yes	1	53	6	25	85
No	191	59	48	88	386
	192	112	54	113	471

ranges from 2 per cent in 'African Beer' to 8 per cent in beer to 12 - 20 per cent in wine and finally to near proof in cane spirits. It will be recalled that these Africans traditionally used fermented drinks of which 'tshwala' was the most common and had a very low alcohol content while being highly nutritious. In recent years local authorities of African administration in the urban areas have started to brew an imitation of this beer commercially. Today in urban residential areas the Africans are prohibited from brewing their own beer — not that this has prevented them from doing so — and they are expected to purchase the commercially brewed 'ijuba', known also as K.B. (Corporation Beer) and by more descriptive terms. In almost any bottle store in Durban a person can purchase 'ijuba' and its sale is not restricted to Africans. Yet, as we can see in Table 5, use of this beverage, designated here as African Beer, is almost exclusively restricted to the Africans.

There are of course a variety of concoctions, most of which are high in alcohol content and unhealthy due to impurity. Such names as 'Skokiaan', 'babaton', 'isiqataviki' (kill-me-quick) and 'shimeyani' are familiar to researchers in urban South Africa and more recently 'gavini' has been added to the list. These concoctions are normally purchased only by members of the lowest socio-economic class. Thus Craig reports on a study of drinking patterns among Africans in the Durban area and finds that 16 persons or 32.0 per cent of patients at the Kwasimama out-patient clinic for African alcoholics preferred such concoctions while only 2 persons or 4.5 per cent of her control group preferred these potions with high alcoholic content (1970: 76). The reverse was also true in that only 22 per cent of the patients preferred African beer while it was the favorite of 80 per cent of the control sample drawn at random.

It will be recalled that in an earlier section the Coloured people were described as a marginal population. This is true genetically, politically and to a smaller degree also culturally. Members of this group who phenotypically resemble Whites frequently "pass for white" (Watson, 1970). While culture and phenotype are obviously independent variables one frequently finds a greater degree of interaction and association when phenotypes resemble each other. This is obviously based, among other things, on parentage and kinship. Thus one also finds that persons who more closely resemble Africans, or who may have been 'classified' more recently as Coloured may in fact have close association and friendship in the African community and may also practice customs more commonly associated with the Africans. A case in point is the use of African Beer. Its use is completely absent among Indians and Whites. Only three persons among the Coloured subjects admitted drinking it but 65 persons, or 28.8 per cent of positive responses among Africans were for African Beer. (It should be pointed out that in Table 5 we can have more responses than our total number of subjects since people may drink and customarily use more than one kind of alcohol.)

Among the Coloured sample we find an 81.1 per cent response favoring liquor while only 3.6 per cent stated that they used wine. In comparison with a national sample of Coloured persons on wine usage this percentage will be exceptionally low.[12] It should be kept in mind that nationally the Coloured population group is concentrated in the Cape Province, long famous for its grapes and wines.

While we researched the age of first cannabis use (since cannabis was the focus of the research project) we unfortunately overlooked ask-

Table 5: Statistical description of the type of alcohol used according to age and ethnic group by a sample of cannabis users

	below 17	18-23	24-29	over 30	Total
African					
African Beer	2	19	27	17	65
Beer	2	23	36	15	76
Wine	0	3	0	1	4
Liquor	2	25	42	12	81
Coloured					
African Beer	2	0	0	1	3
Beer	1	7	3	3	14
Wine	1	2	0	1	4
Liquor	20	44	16	10	90
Indian					
African Beer	0	0	0	0	0
Beer	0	9	1	0	10
Wine	0	11	1	0	12
Liquor	2	17	3	2	24
White					
African Beer	0	0	0	0	0
Beer	3	51	24	5	83
Wine	4	43	13	3	63
Liquor	3	18	8	2	31
Total sample					
African Beer	4	19	27	18	68
Beer	6	90	64	23	183
Wine	5	59	14	5	83
Liquor	27	104	69	26	226
Total	42	272	174	72	

Total sample	African	Coloured	Indian	White	Total
African Beer	65	3	0	0	68
Beer	76	14	10	83	183
Wine	4	4	12	63	83
Liquor	81	90	24	31	226
Total	226	111	46	177	

Table 6: Pattern of first drug use by ethnic group[13]

	African	Coloured	Indian	White	Total
Alcohol	36	34	26	99	195
Cannabis	48	27	9	8	92
Alcohol/cannabis	0	9	3	0	12
Petrol/benzene	0	27	0	0	27
Other drugs	3*	3	0	1	7
Other combinations	0	3	0	0	3
Total	87	103	38	108	336

Summary of distribution:
Mode and variation ratio for:

Africans: Modal category is cannabis — 55.17 per cent
v = .4483

Coloureds: Modal category is alcohol — 33.01 per cent
v = .6699

Indians: Modal category is alcohol — 68.42 per cent
v = .3158

Whites: Modal category is alcohol — 91.66 per cent
v = .0833

ing this question for other drugs or alcohol. We did, however, ask our subjects which drug, and this included cannabis, they used first. There are a number of persons who indicated that they started with both alcohol and cannabis resulting in a combination of these as an alternative category. We combined the whiffers, petrol and benzene, into a single category. Table 6 shows patterns of first drug use by the members of different ethnic groups. The African sample is singular in the high incidence of cannabis while the Coloureds show again the large number of persons who started with whiffers. The Whites in our sample almost all started with alcohol.

In Table 7 the same sample discussed above is analyzed according to the different age categories we distinguished. For the under 17 category, all four ethnic groups, the mode is 45.2 per cent for alcohol as first drug used and 28.6 per cent for petrol and benzene. This latter drug type is however exclusively limited to the Coloured ethnic group. Looking at the same material from the ethnic point of view we find that for the Whites the mode is alcohol with 85.7 per cent and a variation ratio of only 0.1429 while the petrol/benzene category just referred to is the mode for Coloureds with 42.9 per cent and a poor variation ratio of 0.5714. Both Indian and African youths customarily started with alcohol but the numbers in these categories are fairly low.

Turning our attention to the 18-23 year old category, alcohol is the mode for the entire group with 67.4 per cent and a variation ratio of 0.3265. The strongest modal summaries are for Whites and Indians with respectively 88.7 per cent (V=0.1127) and 63.3 per cent (V=0.3667).

Table 7: Pattern of first drug use for each age category by ethnic group

	African	Coloured	Indian	White	Total
Below 17					
Alcohol	5	6	2	6	19
Cannabis	0	5	0	1	6
Alcohol/ cannabis	0	3	0	0	3
Petrol/benzene	0	12	0	0	12
Other drugs	0	1	0	0	1
Other combinations	0	1	0	0	1
Total	5	28	2	7	42
18-23					
Alcohol	3	14	19	63	99
Cannabis	0	12	9	7	28
Alcohol/ cannabis	0	4	2	0	6
Petrol/benzene	0	11	0	0	11
Other drugs	0	1	0	1	2
Other combinations	0	1	0	0	1
Total	3	43	30	71	147
24-29					
Alcohol	19	11	4	24	58
Cannabis	35	5	0	0	40
Alcohol/ cannabis	0	1	0	0	1
Petrol/benzene	0	0	0	0	0
Other drugs	3	1	0	0	4
Other combinations	0	0	0	0	0
Total	57	18	4	24	103
Over 30					
Alcohol	14	3	1	6	24
Cannabis	13	5	0	0	18
Alcohol/ cannabis	0	1	1	0	2
Petrol/benzene	0	4	0	0	4
Other drugs	0	0	0	0	0
Other combinations	0	1	0	0	1
Total	27	14	2	6	49

The Coloureds are nearly trimodal with alcohol (32.6 per cent), cannabis (27.9 per cent), and petrol-benzene (25.6 per cent) thus producing a very poor variation ratio due to the three way distribution. The African responses here are once again too small to be significant.

Among the 24-29 year old category we find a model summary for alcohol with 56.3 per cent but a variation ratio of 0.4369 which is basically due to the 38.8 per cent who started with cannabis. This is the first time that cannabis becomes a significant factor for the African sample where it is the mode at 61.4 per cent. For Whites the mode is alcohol as there is a perfect association. The Coloured sample presents alcohol as the mode at 61.1 per cent and a variation ratio of 0.3889.

The last age category are persons who are over thirty. The total sample here is almost bimodal with alcohol 48.9 per cent and cannabis 36.7 per cent — the variation ratio of 0.5102 being affected by combinations of drugs, especially that of petrol/benzene. The White sample shows again a perfect association with alcohol while the Indian sub-class is too small to count. As before the Coloured sample is almost trimodal between cannabis (35.7 per cent), alcohol (21.4 per cent), and petrol/benzene (28.6 per cent). The African sample, in the clearest bimodal association, favors alcohol (51.8 per cent) as well as cannabis (48.2 per cent).

All of this information is drawn together in Table 8 where age is compared with the first drug used by the total sample of subjects. While there are certain tendencies, e.g. an association between alcohol and 18-23 year olds, or a higher incidence of cannabis use among 24-29 year olds, there is almost no two-way association between age and pattern of first drug use. For the total sample only 26.0 per cent of the comparisons of cases in different beginning use patterns show consistent differences in age group.[14]

In conclusion it should be mentioned that in selecting our subjects it was not always possible to follow the rules of arriving at a truly random sample. For this reason it will be seen that our total Coloured sample is extremely high in the first two age categories, while the White sample is very low at both ends. This is also true for the Indian sample. Our total African sample is perhaps better spread out but the sub-sample drawn by random numbers and used here is very low in the under 17 year old age category. It has been stated that women, particularly African and Indian women, participate much less in the use of consciousness altering substances than males. In this total sample it is really only the White ethnic group that shows a significant spread between the sexes.

Table 8. Pattern of first drug use by age for an inter-ethnic sample of cannabis users

	Below 17	18-23	24-29	Over 30	Total
A. Alcohol	19	99	58	24	200
B. Cannabis	6	28	40	18	92
C. Alcohol/cannabis	3	6	1	2	12
D. Petrol/benzene	12	11	0	4	27
E. Other drugs	1	2	4	0	7
F. Other combinations	1	1	0	1	3
Total	42	147	103	49	341

4. SECULAR RITUAL AND ALTERED STATES OF CONSCIOUSNESS

The material presented in the first half of this study dealt with the ritual of first use and its social significance for participants. This does not deny that the ritual may regularly or frequently be re-enacted, but subsequent rituals do not hold the same significance for the participant who is doing his first consciousness altering activity. These rituals are completely secular in that they define, in an interactional sense, sets of users and networks of access to resources. The drugs which are used will have varying neurophysiological effects, which are beyond the scope of this study, but the decision to use them the first time represents the critical attitudinal change.

The second half of the study presented quantitative material relative to the use of mind altering substances by persons of various age categories belonging to all four ethnic groups. Given the ethnic complexity it is almost impossible to generate hypotheses concerning use of these substances. We are faced by ethnic variation and legal differentiation but a uniform drug law. We also have a variety of consciousness altering substances which may be used by the same ethnic group.

For a long time it was maintained in South Africa, and many who should know better still maintain, that the African uses cannabis because it is inexpensive. Since the African is at the lowest level of the economic scale he would, the argument goes, use cheap alcohol and cannabis rather than the more expensive substances. I have, for some time, argued for a consideration of socio-cultural reasons. The material presented here leans convincingly in the direction of long standing traditions of cannabis use as well as the use of 'tshwala'. These are cultural historical rather than economic reasons. The use of alcohol, as presented in Table 5, does not show a great increase in liquor among older, and thus financially more secure persons. It also does not show a significant number of wine drinkers versus persons using the more expensive liquor. Much the same argument could be presented for the Indian subjects in our sample — and the Indians in Natal are financially fairly well off.

Much the same argument is presented when people state that cannabis is a stepping stone drug leading to 'hard' drugs, or conversely, that many people who use 'hard' drugs, started with cannabis. For this particular sample it is significant that only one person out of our African sample of 196 persons uses any drug (excluding alcohol) in addition to cannabis. While the material is not presented here, we have an age range of persons from 10 years old to about 98, we also have extensive biographical interviews for long-term African cannabis users. In no case do we have any long-term users who have 'progressed to' the use of hard drugs and some of our approximately 1500 African subjects have smoked cannabis for sixty years.

Once again this same picture emerges, though less clearly, for the Indian sample. As in the case of the Africans we have a tradition of cannabis use which is carried into the present. However, among the Indian migrants a century ago there seems to have been a pattern of cannabis and opium use. Now opium use is almost entirely restricted to Whites and Coloureds.

In conclusion, let us return to our first paragraph. The use of drugs is as old as mankind and

> probably began when our ancestors browsed their way through the forests and found that, among the foods they sampled, some produced interesting changes in how they perceived, and how they could accommodate themselves to the world (Aaronson & Osmond, 1971:4).

As patterns of food use and other patterned practices must be seen against socio-cultural and ethno-historical backgrounds, so too must the rituals which accompany various dissociational states. Knowing the cultural setting, substances employed, and reasons for use will assist in defining the 'problem'. Knowledge of such factors will also be invaluable in cases where rehabilitation is necessary.

NOTES

1. Materials presented here are part of a longer investigation: 'A socio-cultural Study of Cannabis in Africa' supported in whole by PHS Research Grant DA 00387, from the National Institute on Drug Abuse. My sincere gratitude is extended for this financial support and other advice and assistance. The statements and conclusions are my own.
2. I had hoped to have at least 120 Indian subjects. Unfortunately, during the critical stages of the research, we lost both Indian research assistants. The persons on whom we do have information, however, are quite representative of the religious, linguistic, and socio-economic groups among the South African Indian population.
3. In the late stages of the research a cognitive questionnaire was administered to matric (twelfth grade) pupils and first year university students at twelve high schools and four institutions for higher learning representing all four ethnic groups. An analysis of this material is not presented here but will appear in du Toit (m.s.).
4. This historical reconstruction of the spread of cannabis use has been presented elsewhere (du Toit, 1974).
5. This does not deny the fact that trances and other dissociational states were prevalent among spirit media or persons who believed that the ancestors were in contact with them. Our reference in the text is to dissociational states induced by the use of some psychopharmacological preparation.
6. I am using 'dissociational state' as a functional term to describe a withdrawal condition consciously aimed for, while this would be included in the more neutral and encompassing 'altered state of consciousness'.
7. In this paper I am not concerned with the justification for treating these drugs the same. The legal implications are our only concern.

8. While conducting biographical interviews in this research, an old Zulu who had been one of the African troops with the South African forces in North Africa during World War II explained: "Frequently before an attack we would be given dagga because the White soldiers said we would all be brave after smoking it". Michael Hoare also suggests that soldiers in the Congo (now Zaire) hostilities had used large quantities of cannabis prior to attacks (1969: 21 and 82). It is also notable that Samora Machel, Frelimo leader and new President of Mozambique, while addressing a gathering of Nampula, recently stated: "... the Portuguese soldiers got their inspiration from alcohol, prostitutes and drugs. To be able to fight they had to drink more and smoke dagga ..." (The Star, June 7, 1975).
9. My sincere appreciation must be extended to Ms. B. B. Edwards for her assistance in the quantitative analysis of this and other research materials.
10. I am indebted to Dr Mervin Moldowan, Department of Pharmacology, University of Florida for assistance in classifying these various drugs according to pharmacological criteria. The major types are as follows:

    A. Central nervous system stimulants:
        1. amphetamines
        2. dexedrine
        3. diet pills
        4. purple hearts
        5. orange hearts
        6. black bomb
    B. Narcotic analgesics:
        1. opium
        2. heroin
        3. morphine
    C. Hypnotics:
        1. mandrax
        2. barbiturates
        3. sleeping pills
    D. Tranquilizers:
        1. 'tranquilizer' unknown
        2. librium
        3. Serepax
    E. Psychoactive compounds:
        1. LSD
        2. hashish
    F. Analgesics:
        1. disperin
    G. Sympathomimetics:
        1. benzedrex
        2. Obex
        3. Ponderax
    H. Anti-cholinergic:
        1. artane
    I. Miscellaneous organic compounds:
        1. benzene
        2. petrol
        3. glue

Obex and Ponderax also have strong anorectic properties and are used as appetite suppressants. As of this date it has been impossible to identify Anacronol.

11. Cannabis Sativa.
12. In an analysis of the consumption of unfortified wines by ethnic group in South Africa, Statsinform (1972:44) found the following:

	% of total for product	% of all liquor consumed by specified race group
Coloureds	46.0	45.4
Africans	31.0	19.1
Europeans	22.5	5.1
Asiatics	.5	2.5

With reference to fortified wines, the picture is only slightly different.

	% of total for product	% of all liquor consumed by race group
Coloureds	37.0	22.4
Africans	33.0	12.4
Europeans	29.0	4.0
Asiatics	1.0	3.8

13. This figure was derived using the statistic θ which measures association between nominally and ordinally scaled data.
14. On the interview schedules three persons in the African sample indicated that they had started by using another drug but did not indicate the kind of drug.

BIBLIOGRAPHY

Aaronson, Bernard & Humphrey Osmond, 1971, Psychedelics. The uses and implications of hallucinogenic drugs. Cambridge: Schenkman Publishing Co.

Balkisson, B. A., 1973, A study of certain personality characteristics of male, urban Indians with drinking problems. Unpublished thesis for the Master of Science Degree, University of Durban — Westville.

Bourguignon, E., 1970, Hallucination and trance: an anthropologist's perspective. In Wolfram Keup (ed.), Origin and mechanisms of hallucinations. New York: Plenum Press.

Bourguignon, E. (ed.), 1973, Religion, altered states of consciousness and social change. Columbus: Ohio State University Press.

Carothers, J. C., 1953, The African mind in health and disease. Monograph Series No. 17, Geneva: World Health Organization.

Chapple, E. D. & C. S. Coon, 1942, Principles of anthropology. New York: Henry Holt & Co.

Craig, Heather D., 1970, The drinking pattern in Kwamashu Bantu Township. Unpublished thesis for Master of Social Science degree, University of Natal.

Dickie-Clark, H. F., 1966, The marginal situation: a sociological study of a Coloured group. New York: Humanities Press.

Durkheim, Emile, 1954, The elementary forms of the religious life. London: George Allen & Unwin Ltd.
du Toit, Brian M., 1971, The Isangoma: an adaptive agent among urban Zulu. Anthropological Quarterly, 44(2).
du Toit, Brian M., 1974, Cannabis sativa in Sub-Saharan Africa. South African Journal of Science, 70(9).
du Toit, Brian M., 1975, Dagga: the history and ethnographic setting of cannabis sativa in southern Africa. In Vera Rubin (ed.), Cannabis and culture. The Hague: Mouton.
du Toit, Brian M., 1976, Continuity and change in cannabis use by Africans in South Africa. Journal of Asian and African Studies, XI(3-4).
du Toit, Brian M., Cannabis use in a sample of South African schools (ms.).
Firth, Raymond, 1972, Verbal and bodily rituals of greeting and parting. In J. S. La Fontaine (ed.), The interpretation of ritual. London: Tavistock.
Gluckman, Max (ed.), 1962, Essays on the ritual of social relations. Manchester: Manchester University Press.
Goody, Jack, 1961, Religion and ritual: the definitional problem. British Journal of Sociology, 12(2).
Hoare, Michael, 1969, Congo mercenary. London: Robert Hale.
Kalant, Oriana Josseau, 1972, Report of the Indian Hemp Drug Commission 1893-94: a critical review. The International Journal of the Addictions, 7(1).
La Fontaine, J. S., 1972, Ritualization of women's life crises in Bugisu. In J. S. La Fontaine (ed.), The interpretation of ritual. London: Tavistock Publications.
Lambo, T. A., 1964, Patterns of psychiatric care in developing African countries. In Ari Kiev (ed.), Magic, faith, and healing. Glencoe: The Free Press.
Laubscher, B. J. F., 1937, Sex, custom and psychopathology. London: Routledge & Kegal Paul.
Leach, E. R., 1968, Ritual. International Encyclopedia of the Social Sciences, 13. New York: MacMillan and Free Press.
Mackay, D., 1948, A background for African psychiatry. East African Medical Journal, 25.
Meer, F., 1969, Portrait of Indian South Africans. Durban: Premier Press.
Millar, L. L., 1972, The problem of alcoholism and drug dependence in the Indian community in relation to the Law on Dependence Producing Substances (M.S., limited distribution).
Opler, Marvin K., 1956, Culture, psychiatry and human values. Springfield: Charles C. Thomas.
Pachai, Bridglal, 1971, The South African Indian question 1860-1971. Cape Town: C. Struik.
Report of the Indian Immigrants Commission, 1887. Pietermaritzburg: P. Davis & Sons Government Contractors.
Samuelson, L. H. (n.d.), Some Zulu customs. London: Church Printing Co.
Statsinform, 1972, A survey of the liquor industry in South Africa.
Turner, Victor W., 1973, Symbols in African ritual. Science, 179(4078).
Venter, J. D., 1965, The drinking pattern of the Coloureds in the Cape and Natal. Pretoria: National Bureau of Educational and Social Research.
Wallace, A. F. C., 1959, Cultural determinants of response to hallucinatory experience. American Medical Association Archives of General Psychiatry.
Watson, Graham, 1970, Passing for White. London: Tavistock Publications.

WILLIAM E. CARTER

5

RITUAL, THE AYMARA, AND THE ROLE OF ALCOHOL IN HUMAN SOCIETY

True to the dualistic tradition that is so much a part of Western society, the secular and the sacred have long been looked upon as polar opposites, separate spheres constantly challenging and buffeting one another With the loss in power of traditional religious institutions has come the implicit acceptance that, for Western man, the sacred has been either totally lost or is swiftly losing its importance.

As the term is used in the popular media, secularization usually refers to the simple fact that post-medieval Western institutions are no longer controlled by the church. Some writers, however, have given it broader meaning. In his book, 'The secular city', Harvey Cox relates it to: 1. the disenchantment of nature, 2. the desacralization of politics, and 3. the deconsecration of values (the abolition of absolutes). He argues that each of these may proceed at different rates, and may be evaluated differently (Cox, 1965:17ff). David Martin goes even further afield. He sees secularization as: 1. institutional: the decline in an ecclesiastical institution's power, wealth, influence, and range of control; 2. ritual: the diminution of religious custom, practices, and rituals; 3. ideas: the adoption of a rationalistic, empirical or skeptical view; and 4. attitudes: the decrease in reverence mystery, deep seriousness, and ultimate concern (Martin, 1969:48-54).

In such approaches, the institutional process is seen as central. The diminution of ritual in Western society is perceived as ranging from challenges to nativity scenes on public property to the emergence of new common ceremonies (e.g., memorials to the Kennedys and King). Seldom is it recognized that an absence of reverence may be nothing more than an alteration in the ritual object. Deats writes, for example:

> There may well be, accompanying the reliance on reason, a reduction of the scope of the mysterious, without any denial of final mystery beyond the reach of reason. And there may well be changes in the objects to which reverence is given and about which persons are deeply serious ... (Deats, 1975:13).

The basic problem in trying to separate the secular from the religious, of course, is that, in most cultures, they are intertwined (Martin, 1969: 3). A case in point is the traditional culture of the rural Aymare of Bolivia and Peru. Westerners, looking at the public behavior characteristic of traditional Aymara, are often shocked by seemingly uncontrolled drunkenness during religious festivities, even in the churchyard and the church itself. The common presumption is that these Indians have no respect for the seriousness of Christian ritual, and that they exploit it as an excuse to carouse. To outsiders, sacred and secular appear so inextricably intertwined in Aymara public ceremonies that some doubt that even the Aymara can distinguish them.

Any object or action, when shared and ritualized, can become a potent binding force. To be viable, every society and social group must have some sort of agreement on these objects and acts, whether codified or not. If truly sacred, such objects and acts enjoy a status that is beyond criticism. Because of their sacred status, to refuse to accept and participate in them can be threatening to the social fabric itself.

Uncritical acceptance, reverence, and compulsive behavior form the foundation of the sacred. The ritual that acts out such acceptance may be either private or public. It may be as simple as touching each lamp post, or stepping on each crack in the sidewalk. It may last a single afternoon or an entire lifetime. But to be accepted as ritual it must involve repetitive action, be kept within limited contexts, reflect basically uncritical acceptance of some value, quality, attitude, or belief, and in some way convey to the individual hope that he will be helped in coping with his situation and in facing life with renewed vigor and confidence.

Among the Aymara, the use of alcohol fits these demands and restrictions remarkably well. Of all the Andean peoples, the Aymara are those who have been most often associated with spectacular drinking and drunkenness. In earlier times they would down enormous quantities of home brewed 'chicha', made with either 'quinoa' or maize. As modern industry has taken hold, they have turned to commercially manufactured beer and, to a much greater extent, distilled cane alcohol from the sugar mills of the lowlands.

In Bolivia three of the most popular brands of this cane alcohol are Guabirá, Venado, and Caymân, all produced in the Department of Santa Cruz. On the 'altiplano' these drinks have almost totally supplanted the traditional quinoa based chicha, a fact that seems at first glance perplexing because of the stringency of the economy. The reason consistently given for the changeover is that, with distilled cane alcohol, total enebriation can be achieved much more rapidly. Such reasoning seems to have a sound basis in fact. Tests run on the alcohol indicate that it is 178 proof, or 89 per cent ethyl alcohol (Heath, 1959). It is so raw that, when undiluted, it sears the lips and can consciously be felt flowing down the lining of the esophagus and hitting the pit of the stomach.

Known simply as 'alcohol', regardless of the brand, the beverage is sold in rectangular cans of one to 16 litres. Characteristically the cans are painted a light purple, are thus readily identifiable, and find their way into other uses once drained of their contents. Aymara society is an incredibly frugal one and makes multiple use of every resource; alcohol is one of the few items which seem to be freely used with no overt thought as to cost or scarcity.

The drinking bouts most commonly observed by outsiders are patronal fiestas, in which characteristically one can see costumed dancers, a variety of musicians, and truly spectacular drinking. Participants become thoroughly intoxicated. They reel, stumble, and collapse utterly. Not infrequently they vomit and return immediately for more drink. To most outsiders such drinking sessions seem orgiastic and totally uncontrolled. But this is a misperception.

At least four things differentiate Aymara alcohol use from its use in Western society. First, it almost always occurs in a social setting. In traditional Aymara society, one is forced to drink specified amounts (usually three shots or three 'fingers' of a bottle) at a command from his host, and a solitary drinker is a rarity indeed. Second, the society produces practically no alcoholics, i.e., people do not usually develop a physical dependency on drink. Third, enebriation to the point of unconsciousness is perceived as a 'good'. A proper host will do everything possible to totally incapacitate his guests through drink. And fourth, alcohol is looked upon as a ritual object and, in traditional society, used only on ritual occasions.

Coca has long been called the 'sacred leaf' of Andean peoples. But, in traditional Aymara society, alcohol is treated as even more sacred. Coca may be used in totally secular contexts, e.g., mundane work in the fields or travelling. It, furthermore, may be used by isolated individuals. Alcohol is reserved for only social occasions, and usually for occasions which deal directly with the sacred. Not a single rite of intensification or rite of passage may begin without offering libations ('ch'alla') to the spirits and then cementing the relationship between the human and the spirit world by sharing the alcohol with all adults present.

How a people can engage in such spectacular drunkenness and still rarely produce alcoholics has long been an intriguing question. In an all-night drinking bout, it is not unusual for a single individual to down as many as 100 'shots'. And such drinking bouts can continue for one day two, three, or as many as eight. Yet between such bouts the populace is totally sober.

The 'good' attached to drunkenness can be illustrated by the cases of two women. The first was attending her niece's wedding. She brought a gift to the first-day reception in the bride's parents' house and so was forced to accept a series of alcoholic 'tokens of appreciation'. Before she was allowed into the houseyard, she had to call from a distance, announcing her gift. The parents of the bride came out to greet her, each poured a libation of alcohol over the gift, and each then proferred three shots of alcohol. Then came the godparents, and finally the bride and groom, accompanied by a band of musicians. The godmother gave three shots, the godfather three, and then the bride and groom each did the same. With each shot that the old lady downed, she grimaced, and after swallowing, spat, commenting on the foulness of the taste. Yet she was not allowed into the houseyard until she had downed 18 shots. Once in the houseyard, she was seated between the bride and groom, and here the process was repeated, with each 'host' insisting that she down three shots, for another total of 18. Only then was she released to 'enjoy' the wedding dance. And as soon as she began to dance, she was accosted by an official 'server' whose principal task was to walk among the guests, a bottle in one hand and a shot glass in another, insisting that they accept the 'love', i.e., drink, of his hosts. Alcohol was obviously disliked by this woman. And there was no indication that a state of drunkenness was any more attractive. Yet she was forced into both by the demands of proper etiquette and ritual.

The extent to which Aymara will go to conform to ritual demands is truly impressive. Rituals are performed for the entry into and exit from all community offices. One man, in his 50's, had occupied every office the community had to offer. Subsequently his wife died, and after a discreet period of time, he took a second wife. He became concerned over the fact that this second wife could share none of his prestige, because she had not accompanied him in the various rituals and posts of leadership he had previously performed for the community. For this reason, and this alone, he volunteered to become 'alcalde escolar', or school overseer for the standard period of one year. At the end of his service, the community came together to fete him, and formally thank him and his new wife for their joint efforts. Out of respect and admiration for the couple, his neighbors and kin plied him and his wife with drink after drink. Finally the wife fell unconscious. Two friends held her upright and carried her around the schoolyard. A third, in token of appreciation, filled a tumbler with alcohol and, while she was still unconscious, poured it down her throat.

Such excesses do sometimes result in physical damage. The most common problem is that of exposure. Men, particularly, fall to sleep along the road on their way home. Not infrequently the exposure during such sleep triggers pneumonia and can lead to death. Insufficient medical work has been done on the population to determine whether there is other organic damage from such drinking. This remains a priority for future research.

Travellers to the small market towns of the altiplano are often under the impression that people engage weekly, or even more frequently, in drunken bouts. The fact that escapes most is that the market towns are centers of assemblage for a large number of rural communities, that many of these communities are themselves subdivided into various zones, and that each zone contains many families. Patronal fiestas come only a few times a year to a given community or zone, and people participate in accord with their relationship to the sponsors of the fiesta. Life crisis ritual and rites of passage are constantly occurring, and they too involve drink, but they do not repeatedly occur to the same people. Thus, although groups of drunk peasants may be found in a given town week after week, seldom are these the same individuals. A given man or woman may drink on only two or three occasions in an entire year.

A listing of major community rituals in the community of Irpa Chico, situated between Viacha and Calamarca on the Bolivian altiplano, will illustrate the types of events at which drinking occurs:

1. January 1, 2, and 3 — New Years and entry of new officials. Outgoing headmen receive gifts from admirers, and offer food and drink to well wishers. Incoming headmen go from house to house offering gifts of coca and alcohol.

2. February 2 — Candle Mass. Each family goes to its 'liwas' (small plots within the 'common' fields of cultivation). They burn copal, offer coca, and pour libations of alcohol for the spirits of the fields. The purpose of the ritual is good harvest.

3. Carnaval. Saturday and Sunday before the beginning of Lent, dance groups are formed in some homes, and dances are performed on nearby pre-Columbian burial mounds. On Monday, Tuesday and Wednesday, each household holds sheep fertility rites, accompanied by heavy drinking and feasting. On Thursday and Friday visits are made to the potato fields, libations of alcohol are offered to the spirits of the fields, and gifts of potatoes and alcohol are given to the 'camana', or guardian of the fields.

4. Palm Sunday. On Palm Sunday, families go to the market town to receive palm branches from the hands of the headmen designate. These branches are a must for ritual protection of the crops. After distributing them, the headmen designate return home with relatives, friends, and admirers. There is much exchange of food and drink.

5. Good Friday. Mass is said in the market town church. Various fiesta sponsors ('prestes pequeños') represent the community, and take the community's religious banners to the church to be blessed. Children who, having turned seven, are moving into middle childhood, accompany the banner into the church, each holding the end of a ribbon tied to the tip of the banner staff. Relatives and friends of each preste pequeño accompany him in the church, spend the night with him in town, and return to the community with him on the morning of Holy Saturday. On the way periodic stops are made to 'refuel', i.e., to drink alcohol. On reaching the community, a fiesta ensues in the preste's house with much food and drink. The celebration ends late on Easter afternoon.

6. June 5, Pentecost — Harvest festival. At nine p.m. on June 4,

the best of the year's agricultural products are selected for veneration. Set in a place of honor within the house on a ritual 'awayu', or carrying cloth, they are given libations of alcohol, wine, and coca, and are adorned with paper streamers and confetti.

Drinking continues until one or two in the morning. At dawn, each family goes to its ripening 'ch'uñu' (freeze dried potato) bed, burns incense to the tubers, and sprinkles them with libations of alcohol. They then butcher a bovine or sheep and sprinkle its blood as a libation over the same tubers. Conspicuous consumption of food and drink fill the rest of the day, interspersed with fertility rites for cattle similar to those held for sheep during Carnaval week.

7. June 16, Corpus Christi. One of the smaller patronal fiestas, celebrated by only a minority of the community. Dance groups are formed, and fiesta sponsors ('prestes') are feted in a fashion similar to that noted for the celebration of Good Friday, Holy Saturday, and Easter.

8. June 25, Saint John's Day. On the evening before, fires are burned in the fields and fortunes told. In those households possessing images of the saint, feasting and drinking fill the night and continue throughout the following day.

9. June 29, Feast of Saint Peter and Paul. This is a major fiesta for the community. A large number of dance groups are formed. Mass is sponsored by various prestes, and enormous quantities of food and drink are consumed.

10. July 16 — day of the Department of La Paz. Although this is a civic holiday, and involves principally marching by school children, some drinking occurs among community leaders.

11. July 25 — Feast of Saint James. Another major fiesta for the community. As with the feast of Saint Peter and Paul, dance groups are formed, mass is sponsored by various prestes, and large amounts of food and drink are consumed. Dance group leaders for the following year are elected.

12. August 2 — Indian day. Civic ceremonies are held in the various schools of the community, led in large part by the students themselves. Drinking rarely although occasionally occurs.

13. August 6 — Bolivian Independence day. Parades are held, after which the parents of school children fete the school teachers with food and drink. After the children and their parents have returned home in late afternoon, the school 'mayors' (i.e. overseers) stay and drink well into the night with the teachers.

14. August 25 — 'San Salvador'. The sponsors, or prestes, for this fiesta consist of individuals who are terminating their successful ascent through the complex hierarchy of socio-religious offices characteristic of the community (cf. Carter, 1964). Three years have passed since they were community headmen, and, following their successful handling of the San Salvador fiesta sponsorship, they will be considered 'pasados',

i.e., men who have reached the highest prestige the community can offer. These individuals ride into the market town on decorated mules, attend the mass they have sponsored, and then return home accompanied by hordes of admirers and well wishers. On returning home they receive well wishers, many of whom come bearing gifts, for three consecutive days in their homes. Huge quantities of food and drink are consumed, for the sponsors have reached their zenith, both in terms of service to and esteem by others.

15. November 1 — All souls day. All night wakes are held in the households in which there has been a death sometime during the previous three years. As in the wakes that accompany death itself, much food and drink is consumed, and offerings of both are left for the spirits of the dead.

16. December 25 — Christmas. Overseers of the schools (school mayors) formally leave their office. In imitation of the traditional ritual accompanying the exit of community headmen on January 1, each overseer is decorated by admirers with ropes of fruit, bread, and beverages, and he and his wife are forced to drink and eat themselves into a stupor.

In addition to these many public rites of intensification, many private rituals exist. Basically they can be divided into two types: rites of crisis and rites of passage.

Of the rites of crisis, the most common are those related to illness, misfortune and healing. Each springs from divination which, in itself, usually involves at least some use of alcohol, both as a libation to the spirits and as a binding mechanism for all present. This dual use of the substance then carries over into the healing rite itself.

Rites of passage are many. They include baptism, the first hair cut, in some families the giving of a protective rosary in early infancy and its withdrawal at about age two, betrothal, marriage, the post-marital giving of bride price and dowry, the first potato planting for the newly married couple, and the roofing of their first house (cf. Carter, 1976). Finally, rounding out each person's life, is funerary ritual (cf. Carter 1968). In all of these, alcohol plays a major role.

A few additional rituals are related to specific roles of leadership in the community. During every moment of their year in office, headmen are expected to carry a small bag of coca and a small bottle (of perfume size) of undiluted alcohol. Whenever they greet a community member, they are expected to invite the individual to partake of both stimulants.

Alcohol also enters the realm of justice. All minor infractions are settled within the community by the council of headmen. Most commonly the sentences are in the form of fines, and the fines consist of specified quantities of alcohol which are consumed on the spot.

Because headmen constantly handle alcohol, they, above all others, run the risk of becoming addicted. Indeed, the very few cases of alcoholism reported in recent community history are all of individuals who became compulsive drinkers during their term of office as headmen.

The obligations headmen must assume not only threaten their normal ability to keep alcohol compartmentalized to ritual acts; they also impose a tremendous financial burden. It is the headmen, for example, who must collect the land tax each August or September. To do so, they must fete their fellow community members as if they were sponsoring a mini-fiesta. This, as other feasts, is given in a spirit of altruistic concern and represents conspicuous giving in its most developed form.

The most common words to accompany the proferring of a shot or bottle of alcohol, regardless of the ritual, is that this represents the host's 'cariño', or love. To reject it is to reject that love.

Individuals will save for years to take on a ritual task. Headmen normally must sell two cows or more to purchase the alcohol demanded of them for their various obligatory ritual acts during their year of office. Since most own no more than three or four such animals, the enormity of their dedication is self-evident.

Even the simplest of ritual tasks, that of dance group leader, follows the pattern. Once a young man decides to assume such a responsibility, and thus begin his climb up the ladder of socio-religious prestige, his first challenge is to recruit other dancers. To do so, he goes to the houses of his relatives and acquaintances and, to each who accepts the invitation to join the dance group, gives half a bottle of undiluted alcohol. One he chooses as choreographer, and to him is given a full bottle, half for his dancing, and half for his choreographic skills.

The basic sacred number for the Aymara is three. In preparing the dance groups for performing, three rehersals are held. In each the dance group leader constantly plies the dancers with food and drink. As do other hosts on other occasions, he usually insists that each drink three shots at a time, or, barring the existence of a shot glass, a three finger-wide depth from the bottle.

As with others who take on fiesta sponsorship, the dance group leader must receive well wishers during the days of the fiesta. He must give food and drink as freely to them as to the dancers themselves. Following the dance, he returns to his home, and there receives more well wishers. In keeping with accepted custom, he and his 'server' do their best to so inebriate his guests that they must pass the night in his house.

Although an analysis has been made of the potency of the undiluted alcohol served frequently on such occasions, none exists for the more common dilutions. Alcohol is mixed with a variety of substances, ranging from water to milk to various fruit juices. With all-night rituals, a common pattern is to begin with weak mixtures, and to gradually work up to the pure liquid, just as it comes from the can. Given our present state of knowledge, it is impossible to say exactly how much of this 'pure' alcohol is drunk, on the average, by any one individual. However, it is obvious that, even when taken in a diluted fashion, the amount is impressive. In one betrothal party lasting a single night, for example, the two extended families (consisting of 14 people) finished off a 16 liter can.

It was earlier stated that, among the Aymara, alcohol is used as a binding mechanism between people. It lowers normal reticence, breaks down hostilities, and gives people who live in a very harsh and demanding environment a chance to forget their troubles, be it for ever so little. In a clearly imitative fashion, alcohol binds individuals to the spirit world. Offerings to the spirits placate their capriciousness and render them less threatening. Even corpses are outfitted with their bottle of alcohol so that they may have strength in their journey to the next world.

As the traditional Aymara use alcohol, it would appear to be at the polar extreme from the secular objects and acts so overwhelming in contemporary Western society. Alcohol among these people has a power and influence far beyond its simple potency as an intoxicant. It is surrounded by reverence, mystery, deep seriousness, and ultimate concern. The fact that it forms an integral part of public rituals, even Christian rituals, does not mean that the Aymara do not take these rituals seriously, or that they have secularized them. Rather, drink and drunkenness are themselves elevated to the level of the sacred, are looked upon as one of the greatest goods available to mankind, and are actively courted in all major ritual acts.

The true outcasts in such a society are those who refuse to imbibe, who refuse to accept the love, or 'cariño' of their hosts. For most, drink and drunkenness enjoy a status that is beyond criticism, and therefore that is an integral part of the sacred.

No single explanation could suffice for the fact that, in spite of the spectacularity of drinking sessions among traditional Aymara groups, only rarely does an individual become a compulsive drinker or an alcoholic. Two variables must lie at the core of whatever explanation is given, however. First, all such drinking is social. Bars do not exist, and solitary drinking is practically unheard of. Second, drink in traditional Aymara society symbolizes the religious. Without it, one cannot enter the world of the sacred. The counterpart of this attitude, of course, is that one does not handle drink lightly. It is reserved for those occasions of ultimate seriousness.

The way in which the traditional Aymara handle drink varies drastically from the way it is handled in Western society. Their use of alcohol teaches us, first of all, that we cannot hastily interpret the meaning of what, at first glace, appears to be an outrageous act in a strange society. Second, it teaches us that occasional drunkenness to the limits of excess should not and cannot be equated with debauchery. And third, it teaches us that such drunkenness can and does occur repeatedly without becoming physically addictive. The Aymara experience clearly shows that, like other drugs, alcohol, its use, and its effects, are conditioned more by the cultural setting than by the nature of the intoxicant itself. The Aymara propel a substance that is normally associated with the secular in Western society into the realm of the sacred. One wonders how often this occurs in other societies without our perceiving it. Perhaps, even in our own, this is occurring. The sacred, rather than withering away, may simply be changing its focus. We may be witnessing alterations in the objects to which reverence is given. And alcohol, as consumed, shared, and ritualized in our cocktail parties and bars, may be on its way to becoming one of these revered objects. The frequency of its use in such social settings mitigates whatever 'protection' increased sacralization may promise, however. Propelling alcohol into the realm of the sacred can soften the substance's addictive powers only when sacred acts are finite and limited in number.

BIBLIOGRAPHY

Carter, William E., 1964, Aymara communities and the Bolivian agrarian reform. Gainesville, Florida: The University of Florida Press.

Carter, William E., 1968, Secular reinforcement in Aymara death ritual. American Anthropologist (Menasha, Wisconsin), 70(2):238-263.

Carter, William E., in press, Trial marriage in the Andes? In Ralph Bolton & Enrique Mayer (eds.), Andean kinship and marriage. American Anthropological Association Special Publication.

Cox, Harvey, 1965, The secular city. New York: Macmillan.

Deats, Paul K., 1975, The sacred criticism of secular culture. Nexus. Boston: Boston University School of Theology.

Heath, Dwight, 1959, Drinking patterns of the Bolivian Camba. Quarterly Journal of Studies on Alcohol (New Haven, Connecticut), 19:431-508.

Martin, David A., 1969, The religious and the secular. London: Routledge & Kegan Paul.

WAYNE M. HARDING & NORMAN E. ZINBERG, M.D.*

6

THE EFFECTIVENESS OF THE SUBCULTURE IN DEVELOPING RITUALS AND SOCIAL SANCTIONS FOR CONTROLLED DRUG USE[1]

In the United States, social and legal taboos against the nonmedical use of illicit drugs are reinforced by the prevailing view that these drugs are almost animately pernicious. According to this view, marihuana, LSD, cocaine, heroin, and other ilicit drugs are so overpowering and/ or so dangerous that their continued use inevitably leads to drug abuse. The physiological and psychological damage evidenced by the most serious abusers of illicit drugs is regularly invoked as proof of this "pharmacomythology" (Szasz, 1975).

There is nothing in the pharmacology of these drugs, however, that precludes the possibility that they can be used without being abused. Our study of controlled drug use, sposored by The Drug Abuse Council, Inc., has located users of marihuana, psychedelics, and opiates who, like most alcohol users, manage to maintain regular non-compulsive use of these drugs. Analysis of longitudinal interview data indicates that this 'controlled' use is chiefly supported by emerging subcultural drug-using rituals and social sanctions. These rituals and social sanctions provide what the larger culture does not: instruction in and reinforcement for maintaining patterns of illicit drug use which do not interfere with ordinary functioning and methods for use which minimize untoward drug effects.

In this article we discuss these findings and the related work of other researchers. We also argue that existing subcultural rituals and social sanctions, elaborated and endorsed by the mainstream culture, could be a more humane and perhaps more effective means of preventing drug abuse than legal prohibition.

Serious consideration of such alternatives is especially timely given the recent actions of some states to significantly reduce the legal penalties surrounding the use of marihuana. It appears that these reductions have been prompted by a growing realization that our costly social policy

* Wayne Harding is a Research Associate at The Cambridge Hospital. Norman Zinberg is a Faculty Member of Harvard Medical School at The Cambridge Hospital and of The Boston Psychoanalytic Institute.

has not succeeded in halting marihuana use by a large number of Americans. Thus far, however, public debate over liberalization of drug laws has not taken into account changes in drug-using style.

DEFINITION OF TERMS

As used here, 'ritual' refers to the stylized, prescribed behavior surrounding the use of a drug. This behavior may include methods of procuring and administering the drug, selection of physical and social settings for use, activities undertaken after the drug has been administered, and methods of preventing untoward drug effects.

'Social sanctions' refers to the norms regarding how or whether a particular drug should be used. Social sanctions include both the informal and often unspoken values or rules of conduct shared by a group, and the formal laws and policies regulating drug use.[2] These two aspects of social sanctions are not always consonant. Laws prohibiting use of illicit drugs may reflect the values of the majority of Americans but are often at odds with the values of drug users. Various segments of society thus observe quite different social sanctions (and rituals) although each segment is cognizant of and influenced by the other's. The relationship among the rituals and social sanctions of controlled illicit drug users, of compulsive users, and of the mainstream culture is a focus of concern in later portions of this paper.

Our use of the terms 'ritual' and 'social sanction' differs from the classic use of the terms 'ritual' and 'ritual belief' in anthropology. The distinction between drug-using rituals and social sanctions is one of behavior versus beliefs, or practice versus dogma. In anthropology, terms such as 'ritual beliefs' and 'ceremonial beliefs' are used instead of 'social sanctions' (Leach, 1968). We prefer 'social sanctions' for two reasons. First, this term emphasizes that beliefs are socially derived and reinforced. Second, 'social sanctions' conveys more clearly than 'ritual beliefs' the sense that behavior and belief are separable concepts. While it is true that rituals and ritual beliefs are intimately related, and sometimes virtually indistinguishable, we have found that different drug users (heroin addicts versus controlled heroin users, for example) may share very similar drug-using rituals, yet subscribe to dichotomous social sanctions. In other words, social sanctions can be used to predict the type of drug use when rituals cannot.

The terms 'rituals' and 'ritual beliefs' have been applied most frequently to magical or religious phenomena. Goody and others have included secular events (e.g., civil marriage ceremony) under the rubric of ritual, but reserve the term to describe behavior in which "the relationship between means and ends is not intrinsic; is either rational or non-rational" (Goody, 1961). What is usually excluded is any behavior which "is technical or recreational" (Gluckman, 1962).

Our use of ritual and social sanction violates this tradition in two distinct ways. First, we are applying these terms to drug use whether the goal of the user is recreation, improved mental or physical performance, or religious experience.[3] Second, drug-using rituals and social sanctions include both rational and nonrational elements. The intravenous injection of heroin is causally related to the subsequent **high** while **booting** (drawing of blood back into the syringe and reinjecting one or more times) is not, although users may believe that it is.

Our departure from the more restricted meaning of ritual is not without precedent among anthropologists. Klauser (1964), for example, discussed the cocktail party as a ritual. It is worth explaining, however, why the concept 'ritual', even in modified form, is so aptly applied to drug use.

Within very broad limits, the objective and subjective effects of a psychoactive drug depend as much on how the drug is used and the expectations of the user as on its chemical properties. Booting does increase some heroin users' sense of euphoria. A placebo can alleviate pain as effectively as morphine provided the user believes he is receiving an analgesic. Tobacco acts as a powerful hallucinogen in some Amazonian tribes where it is used infrequently in high doses (Weil, 1972). These are but a few examples of the mutability of drug effect which can be attributed to the discrete influence of rituals and social sanctions, whether rational or nonrational, on the drug user. Szasz (1975) similarly justifies applying the term to drug use because it reveals the enormous range in the consequences of that use which are otherwise hidden by a strictly pharmacological perspective:

> Perhaps because of all the major modern nations, the United States is the least tradition bound, Americans are most prone to misapprehend and misinterpret ritual as something else: the result is that we mistake magic for medicine, and confuse ceremonial effect with chemical cause.

Finally, in this paper we are mainly interested in drug-using rituals and social sanctions of a specific kind: those which foster controlled drug use. Drinking muscatel from a bag-wrapped bottle while squatting in a doorway, or soliciting psychedelics from strangers on a street corner is not a controlling ritual. The positive social status attached to the ability to withstand extraordinarily high doses of LSD, the risk involved in **getting loaded** on barbiturates and alcohol, or the size of one's heroin habit does not constitute a controlling social sanction. In the following section we outline the nature of social sanctions and rituals which do promote control, using alcohol as an example. This discussion will provide a basis from which to examine the existing subcultural social sanctions and rituals which facilitate the controlled use of illicit drugs and inhibit their abuse.

RITUALS, SOCIAL SANCTIONS, AND CONTROLLED ALCOHOL USE

Although alcohol is a powerful and addictive psychoactive drug which can produce profound physiological and psychological damage, the vast majority of Americans who drink alcohol manage to control it. There are an estimated 105 million drinkers in the United States compared to some 8 million alcoholics (New York Times, April 9, 1973). Widespread controlled alcohol use can be understood in terms of culturally based rituals and social sanctions which pattern the way the drug is used.

Alcohol-using rituals define appropriate use by limiting consumption to specific occasions or circumstances. Having a highball before dinner, wine with a meal, a few drinks at a cocktail party, a beer with the boys after work, or a drink at a business luncheon are examples. Positive social sanctions permit and even encourage moderate use of the drug: one need only consider the occasions when a drink is offered to appreciate how well alcohol is integrated into the culture as an approved social intoxicant. This social acceptance of alcohol is paralleled by the minimal legal restrictions on its consumption, and by the negative sanctions which condemn promiscuous use and drunkenness. "Know your limit," "Don't drink and drive," "Don't mix drinks," and "Never drink before noon" are familiar proscriptions.

The internalization of these social sanctions and rituals begins in early childhood. The child sees his parents and other adults drinking. He learns the possibilities of excess and the varieties of acceptable drinking patterns from newspapers, movies, magazines, and television. As he matures, he develops a more unconscious than conscious sense that alcohol use can be pleasant, controlled, and socially approved. In some cases, this socialization process is more direct — children sip wine at religious rituals and celebrations, or taste their parents' drinks. Many authorities believe that a gradual and careful early introduction to alcohol by parents contributes to restrained adult use.[4]

Many adolescents drink without parental permission, and some test the wisdom of the social sanctions and rituals with which they are already familiar by getting drunk and nauseated. However, the central issue of this testing is not so much how to drink as it is how long the adolescent must defer approved social drinking. Neither the adolescent nor his parents have much fear that occasional undercover experimentation will seriously or permanently disrupt social relationships and performance at school or work. Throughout this period of early use, the adolescent has numerous adult role models for controlled use and he can easily find friends who share his interest in drinking as well as his resolve to avoid compulsive use.

At some point the young user receives direct or tacit approval for drinking from parents and other significant adults, marking the end of family-centered socialization in the use of alcohol.[5] As the user begins to drink in public, he melds the general culture's rituals and social sanctions and his previous learning into an individualized but socially acceptable pattern of alcohol use. Social reinforcement for controlled use continues throughout adult life.

Obviously the influence of rituals and social sanctions on the alcohol user is partial and imperfect. Other variables — social forces, personality factors, and perhaps genetic differences — also influence how groups and individuals use the drug. The social sanctions and rituals associated with controlled use are not uniformly distributed in the culture. Some ethnic groups (e.g., the Irish) tend to lack strong sanctions against drunkenness and have a correspondingly high incidence of alcoholism (Wilkinson, 1970). Furthermore, even when functioning rituals and social sanctions are available, family-centered socialization may break down. Nonetheless, prevailing rituals and social sanctions exert a discernible, and crucial, moderating influence over the way most Americans use alcohol.

The importance of such rituals and social sanctions has been dramatized by the disastrous effects of the introduction of alcohol to societies which lacked them. American Indian tribes demonstrate longstanding, controlled, highly ritualized use of naturally occurring psychoactive plants such as jimson weed and peyote (LaBarré, 1938). The Indians' legendary susceptibility to alcoholism stems essentially from a lack of similar cultural conventions for the use of the **white man's** drug. Because the Indian has rejected and has been denied full membership in American society, his inculturation in alcohol-using rituals and social sanctions has been retarded. Consequently, alcoholism persists among Indians and the "consequences of alcohol use are frequently deep inebriation, rather than courtly pleasantries" (Freedman, 1974). Wilkinson (1970) reports that when the Eskimos of Frobisher Bay, Baffin Island, were first granted legal permission to drink, their lack of previous cultural experience and guidelines for alcohol use resulted in pronounced abuse.

A similar problem exists for Americans who use illicit drugs. It is not at all surprising that so many of these people wind up as compulsive users. There are virtually no socially accepted models for the controlled use of these drugs, no positive cross-generational education in how to use them, and no reinforcement or assistance in moderate use (Abrams, 1972).[6] The mainstream culture not only fails to assist controlled, illicit drug use, it actively discriminates against it. Any and all use of illicit drugs is prohibited. Persons who use these drugs are regarded as deviant: either as **sick** and in need of counseling and rehabilitation, or as criminal and deserving of punishment. It is clear that use and abuse of illicit drugs must be understood from a sociocultural as well as a pharmacological perspective.

By and large, the research literature reflects the reigning cultural outlook on illicit drug use in that it fails to differentiate between use and abuse. One reviewer of 35 recent studies states that their most serious flaw is that "they have lumped together all drug users without considering the extent of their use" (Heller, 1972).

Patterns of drug abuse such as heroin addiction have been singled out for intensive study, but there has been little effort to delineate patterns of use lying between the extremes of abstinence and abuse or compulsive use. The lack of a definite typology for drug-using behavior bespeaks the continuing and pervasive tendency to confound quite different patterns of drug consumption.

The terms in the literature which are closest to controlled use are 'chipping', 'occasional use', 'experimenting', and 'tasting'. 'Chipping' and 'occasional use' are usually associated with heroin and the opiates. 'Taster' (Kaplan, 1971) and 'experimenter' (Keniston, 1968-69) have been specifically applied to marihuana and psychedelic users. All these terms refer to irregular, nonaddictive, or minimally abusive drug use, but do not necessarily connote the elements of moderation, regularity, stability, and nonabuse which we mean by controlled use.

A computer search of the MEDLINE file[7] covering a 47-month period (January 1969 through November 1972) produced no articles specifically concerning occasional use of any drug. An informal search for mention of occasional use, however, yielded several allusions to occasional use. Jordan Scher (1961, 1966) mentions the existence of controlled heroin use in work done through the Cook County Narcotic Project. Isador Chein et al. (1964) note the existence of "long continued, nonaddictive heroin users." Howard Becker (1963) discusses occasional marihuana use as a stage preceding regular use during which "the individual smokes sporadically and irregularly" because he has not yet established a reliable source for the drug. W.H.Dobbs (1971) warns that not all applicants to methadone programs who are using heroin may be drug dependent. John Newmeyer (1974) found some heroin users who, he feels, should not be regarded as representative of a junkie population because they "could sample heroin without becoming addicted."

The focus in each of these sources is more on regular than controlled use, and little importance is attached to different using patterns. The authors do not seriously consider regular controlled use as a stable use pattern for a significant number of people.

To our knowledge only one published study (Douglas Powell, 1973), also sponsored by The Drug Abuse Council, Inc., focuses specifically on occasional drug use or occasional users. Powell interviewed subjects who had been occasional users of heroin for at least three years without becoming physically addicted. Many of the using patterns described in Powell's report, however, appear so unstable or so damaging that they lie outside the patterns of controlled use we are investigating. Still, Powell's study supports our efforts in that he established the existence of occasional (if not controlled) heroin users and he found that such users "are responsive to research and can be studied reliably with relatively simple techniques."

METHODS OF THE DAC STUDY

The major goals of the Drug Abuse Council study are:
1. to locate controlled users of marihuana, psychedelics, and opiates;
2. to describe such users and their various patterns of use; and
3. to identify factors which stabilize and destabilize controlled use.

Potential subjects were initially solicited through universities, advertisements in the underground press, and a variety of social service agencies including halfway houses, drug treatment programs, and counseling centers. Once underway, we found, as Powell did, that after completing the screening/interview procedure, subjects were often willing to refer drug-using friends and acquaintances to the project. Six indigenous data gatherers (i.e. members of the drug-using subculture) were recruited to assist in locating and interviewing subjects.[8]

The following are the minimum criteria developed for participation in the project.

1. Subjects had to have used marihuana, a psychedelic, or an opiate for at least one year.

2. Subjects had to be willing to participate in followup interviews.

3. A subject had to have used the drug frequently enough to be considered a regular user, but not so frequently that he was physically addicted to it (in the case of opiates) or that his level of use was likely to interfere with effective personal and social functioning. No precise cutoff points for frequency of use were established. In practice, a marihuana user who had used only a dozen times in the previous year was not selected because his use seemed too infrequent to be regarded as regular, and a weekly user of psychedelics was not selected because such frequency suggested a possibly abusive drug-using pattern.

4. When subjects were polydrug users, all of the drugs used (including alcohol) had to be used rather than abused. A subject who was a moderate bi-weekly heroin user, but who was physiologically addicted to barbiturates, was not eligible to participate.

Interviews lasted from one and one half hours to two hours or more. Subjects were paid approximately $10 per interview. A flexible interview schedule was adopted to allow the interviewers to pursue interesting issues as they arose. For each subject data were gathered on his history of drug use (including alcohol); his relations to work and school, as well as to family and mates; his relations with drug-using and non-drug-using peers; his physical health and emotional stability;

details of drug-using situations; and basic demographic variables such as age, years of schooling, and social class.

Profile of the sample

For approximately two years interview data have been gathered on 105 controlled users.[9] The sample consists of 66 white males, 24 white females, 9 black males, and 6 black females. Subjects range in age from 14 to 70 years with most in the 18- to 25-year-old age bracket. Eighty-seven interviewees demonstrate controlled use of marihuana, 42 have used psychedelics in a controlled way, and 46 are controlled opiate users (categories overlap). Followup interviews have been conducted and are still in progress.

We found that the 105 controlled users can be distinguished from compulsive users along several dimensions. Subjects maintain ties to institutions like work or school, and regular social relationships with non-drug users as well as users. Drug use is important to these subjects but is only one of many other activities, relegated to leisure time. Most subjects are deviant only by virtue of their drug use. Some have a history of criminal activity or school disciplinary problems, which does not generally overlap their controlled use of a drug. No subjects manifest physiological or psychological impairment as a result of controlled use.

Our data contradict the notion that the period of controlled use is a brief transition stage ending in abuse or abstinence. Subjects with relatively short histories of controlled use — slightly over one year, for example — are included in the sample to clarify the manner in which controlled use is first established. Long-term followup will reveal how stable these subjects' patterns of use are. The majority of subjects, however, have been controlled users for several years, and some have maintained controlled use for as long as ten years.

RITUALS, SOCIAL SANCTIONS, AND CONTROLLED MARIHUANA, PSYCHEDELIC, AND OPIATE USE

Having outlined our methods and profiled the sample, we will confine ourselves here to a discussion of preliminary findings on the relation between rituals, social sanctions, and controlled use.[10] The most striking feature of the DAC subjects is that they have acquired and adhere to rituals and social sanctions which provide a structure and a mythology for maintaining controlled use and avoiding untoward drug effects.[11]

Acquisition of rituals and social sanctions took place over the course of subjects' illicit drug-using careers. The details of this process varied among subjects: some had been controlled users from the outset of their drug-taking; others had been through one or more periods of compulsive use before firmly establishing control. Virtually all subjects, however, required the assistance of other users to construct appropriate rituals and social sanctions out of the folklore and practices of the diverse subculture of drug takers.

It is this association (often fortuitous) with one or more controlled users which provides the necessary reinforcement for avoiding compulsive use. The using group redefines what is a highly deviant activity in the eyes of the larger culture, as an acceptable social behavior within the group. It reifies social sanctions and rituals and institutionalizes controlled use. This is consistent with Jock Young's (1971) observations of drug use in London where he found that some groups "contain lore of administration, dosage, and use which tend to keep . . . lack of control in check, plus of course, informal sanctions against the person who goes beyond these bounds."

All but two of the DAC subjects have been connected to a controlled using group. Although subjects sometimes use drugs alone, upwards of 80 per cent of their use takes place with others. Use in the company of drug abusers is rare. Controlled heroin users, for example, tend to limit their contact with heroin addicts to those occasions when it is necessary to obtain their drug and to decline invitations to **shoot up** with their addict-suppliers.

While association with controlled drug-using groups is the primary source of controlling rituals and social sanctions for illicit drug use, it appears that the alcohol education process may be a secondary source, especially in the case of marihuana use. Subjects often draw pointed comparisons between social drinking and their use of illicit drugs. Younger subjects apply the same language — phrases like **getting high** and **getting off** — to both alcohol and illicit drugs. Subjects describe social gatherings where both alcohol and marihuana are available and where an individual's preference for one of these drugs over the other is interpreted as a matter of personal choice rather than as a symbolic ideological statement about being in or out of the drug culture. Some subjects treat alcohol and marihuana in much the same way. John L., 26, is enrolled full time in a university and holds down a part-time job. When he returns home he usually has a drink or a joint before dinner, depending, he explains, on his mood and his plans for the remainder of the evening. It seems then that controlled users adapt alcohol-using rituals and social sanctions to their use of illicit drugs.

Taken as a whole the rituals and social sanctions toward controlled illicit drug use have several major features:

1. They define and approve controlled use and condemn compulsive use.
2. They limit use to physical and social settings conducive to a positive drug experience.
3. They incorporate the principle that use should be kept infrequent

enough to avoid dependence/addiction and to maximize the desired drug effect.

4. They identify potential untoward drug effects and prescribe relevant precautions to be taken before and during use.

5. They assist the user in interpreting and controlling his drug high.

Rituals and social sanctions vary with the pharmacology of the three drug types we are investigating — marihuana, psychedelics, and opiates — and with the acceptability of these drugs within and outside the drug subculture. Therefore, the following, more detailed discussion of rituals and social sanctions proceeds by drug type.

Marihuana

Marihuana use is less ritualized than psychedelic and opiate use. Subjects use the drug in a wide range of settings and circumstances: before going to a movie, at a party, while watching television, or during a walk in the woods. Controlled users do not usually come together specifically to take marihuana; they meet to socialize and the drug is sometimes taken as an adjunct to the occasion. Marihuana is also more likely to be used alone than the psychedelics or opiates.

This flexibility in marihuana rituals is in part due to the pharmacology of the drug. Marihuana is a relatively mild and short-acting intoxicant. Our subjects, as experienced users, find no difficulty in controlling the drug high,[12] and they are able to function normally if that becomes necessary. The high state, therefore, is compatible with a variety of public and private settings.[13] A marihuana high is also easily arranged, requiring neither the apparatus to inject an opiate nor the planning to accommodate a 6- to 8-hour psychedelic high.

Flexibility in marihuana rituals can also be explained in terms of the drug's status. The expanding number of marihuana users as well as the growing acceptance of the drug among users and non-users alike[14] has created an environment in which rigid external controls in the form of rituals are no longer necessary. They have been supplanted by controlling social sanctions which are less specific and can be adapted to various using circumstances. DAC subjects 25 years old and over who began using marihuana in the early to mid-1960's describe the more marked ritualization of that period. They recall with nostalgia and humor the dimly lit room, locked doors, music, candles, incense, people sitting in a circle on the floor, and one joint passed ceremoniously around the circle. They now regard this behavior as quaint and unnecessary. As the number of intermittent marihuana users has risen to some 8 million Americans and the number who have tried the drug to 26 million (Boston Globe, 1974), marihuana use has lost much of its deviant character. Concurrently, social sanctions for controlled use have been strengthened and have become available throughout most of the using subculture.

Under these conditions considerable learning about controlled use can take place before use actually begins. The choice of whether or not to use marihuana has become a reality for American adolescents, and most are well aware before making that choice that marihuana does not cause people to go crazy or to fall apart. Younger DAC subjects (18 to 20 years) had known of teachers in their high schools who used marihuana. Many had older siblings who they knew used the drug. These subjects had also acquired a sense of what marihuana was like from friends, the underground press, popular music, novels, and other sources. Their first few experiences with marihuana were usually ritualized affairs with one or more newcomers introduced to the drug by a more experienced user in a secure setting.[15] The experienced users typically provided guidance, demonstrated how best to smoke the drug, and soothed newcomers' lingering fears. Very quickly, though, neophyte users moved beyond these structured situations and began the process of adapting use to a variety of social settings. Most were able to locate friends with whom to use the drug and with whom they also shared non-drug-centered interests. The lack of highly specific rituals should not, therefore, be construed as evidence that controlled users are reckless in the way they use marihuana. Rather, the rituals that earlier served as rigid and external controls have been replaced over the last decade by more general but equally effective social sanctions. Due to growing familiarity with every aspect of marihuana use, these sanctions, like those of alcohol, are internalized; the rituals developed to support these sanctions no longer need to be so closely adhered to. Interviews with subjects reveal how these social sanctions operate to ensure control.

Subjects describe marihuana as a relatively innocuous drug, easily controlled, and difficult to abuse.[16] Some expressed genuine surprise when we asked if they had ever had any difficulty in maintaining controlled use. Subjects are not, however, messianic about marihuana. They recognize its potential for abuse and offer guidelines for sensible use:

> In spite of all the rationalizations about how good dope is, I don't see that I have to have a reason for getting high every time but yet getting high consistently without a reason for it seems to be a reason to sort of check things out with yourself.

Another subject comments that if marihuana is used too much the quality of the high declines and when this happens one should stop for a while and then return to a pattern of more infrequent use. Subjects generally subscribe to the ethic that they should not be high at work or at school. Susan S. works as a housekeeper several days a week. She explains that although she can clean when she is **stoned**, she prefers to restrict her drug use to leisure time.

Controlled users also express the idea that too much marihuana should not be used at any one time. There are two reasons cited for this:

1. to avoid transient but unpleasant panic reactions or **paranoia**, and
2. to keep the high controllable so that other activities can be better enjoyed.

While passing a joint around a group is no longer **de rigueur**, it still serves on many occasions to assist the process of adjusting the intensity of the high. It allows time to pass between each inhalation during which the user can monitor his own degree of intoxication. Several subjects state that when using alone or with one or two other people, they stop after several **tokes** to let the high catch up with them and then decide whether they want more. One subject comments that this is an especially sensible way to proceed when trying out a new batch of marihuana.

Psychedelics

Psychedelics include a wide range of substances that vary both in potency and duration of effect: LSD, mescaline, peyote, psilocybin, MDA, DMT, and others. The illicit status of these drugs creates a major problem for the user; he cannot be certain what is in the drug he is sold.[17] What is presumed to be mescaline may be LSD. It may be adulterated with PCP, amphetamines, and other substances — and its dosage can only be guessed at. Unlike the marihuana high, the psychedelic high[18] usually lasts for several hours. It is an intensive though not uncontrollable experience, characterized by perceptual changes, sometimes of a hallucinatory or illusory nature. The risk of a **bad trip** is always present and to some degree increased by the lack of quality control over the drug. For these and other reasons, psychedelics are regarded as **real**, i.e., dangerous, drugs within the drug subculture. They do not have the widespread appeal of marihuana nor are they treated casually. Most of the rituals and social sanctions related to the psychedelics deal with making the drug experience as safe as possible for the user.

For the subjects, psychedelic use is almost invariably a drug-centered, group activity. Subjects talk about having others with them who can be relied upon to help cope with a bad trip or unforeseen events as a requisite for safe tripping: "I have to do it . . . with someone that I really know well, that I really trust, and there are some people like that." People who are less intimately acquainted are sometimes included in the group but if so, the trip is commonly preceded by a discussion in which everyone tries to get comfortable with one another, to determine who may need extra help or attention, and to establish ground rules for the trip. During this preliminary discussion, an experienced user may be assigned to act as a guide for a more inexperienced or uneasy user. Group members may decide to forbid wandering off from the group without letting someone know because it causes people to worry, and worrying is felt to be detrimental to a positive drug experience.

Subjects agree that planning the trip is an important matter, even when participants have taken the drug together before and feel quite close to one another. The need for structure varies, but pre-trip planning includes issues such as: what foods or beverages to take along, what activities to engage in during the trip, whether thorazine or niacin should be available in case of a bad trip, or whether **talking people down** is preferable to medicating them. This planning reaffirms the participants' sense of shared intentions and strengthens their capability to control the drug high.

Subjects are adamant about using psychedelics in a proper setting — **a good place.** For many this means tripping in a relatively secluded spot in the country. What seems important, however, is that the space is secure and comfortable. A city tripper said, "I'll take a walk outside but it'll always be with the notion that I can come back to this kind of sanctuary for myself in the house, and so it's no threat." This subject and many others expressed surprise and some disdain for users who violated the principle that psychedelic use is a taxing experience that should be confined to special settings:

> I'm amazed that ... I was living last year with a dude who's 17 years old and is from the West Coast. He was telling me that when he was going to junior high school he would just drop acid in the morning and go to school, which completely weirded me out ... and just could ride with any kind of horrible thing ... Amazing.

Another social sanction/ritual which subjects observe is the need to be internally prepared for psychedelic use. One subject describes this as "making peace with the public reality ... mentally putting your house, your affairs, in order, you know, like, what's the Zen thing ... emptying out the teacup first." Others talk simply about needing to be in a "good mood" and needing "energy" to undertake the experience. Some subjects appear to ritualize this internal process by tidying up the space in which they are going to use the drug.

All the conventions described above represent attempts to ensure a **good** trip and prevent a **bad** one. We now turn to the issue of how rituals and social sanctions may inhibit compulsive psychedelic use.

Subjects repeatedly advocate using psychedelics at no less than two-week intervals. In practice, their use is far less frequent than this — less than once a month is the most typical using pattern and, with time, use consistently becomes even less frequent. Avoidance of compulsive use, however, is probably not so much the consequence of negative sanctions as it is the result of a combination of two other factors:

1. the positive value controlled users attach to the consciousness-altering properties of psychedelics, and
2. the fact that tolerance to these consciousness-altering properties goes up very rapidly as use becomes frequent.

Our subjects who are interested in experiencing precisely these effects find that too frequent use of the drug is counterproductive.

Some psychedelic users who are not interested in the consciousness-changing qualities of these drugs may become compulsive users. For them, it is the **speedy**, stimulating effects of psychedelics that are appealing[19] — effects which are enhanced with larger, more frequent doses of the drug. Although we have little direct evidence to support it, we would guess that this kind of compulsive psychedelic user is associated with those groups in the subculture which negatively value consciousness change or do not recognize it as a primary drug effect.

By comparing older and younger subjects we have identified some shifts in psychedelic-using rituals and social sanctions. Subjects who began use in the mid-sixties share a sense that psychedelics should be used for "personal growth" rather than recreational purposes. They discuss tripping as an activity which is undertaken to accomplish a worthy goal — to learn more about oneself, to grow intellectually, to transcend ordinary perceptual boundaries, and so on. However, subjects who began use in the past five years have broadened their reasons for using psychedelics to encompass plainly recreational goals.

Younger subjects may trip for a highly rationalized purpose but they are equally inclined to trip simply to enjoy the high state. This trend is difficult to interpret and we have yet to make final judgments. We speculate, however, that the expanded goals of psychedelic users indicate a growing familiarity with psychedelics and less guilt about their use. Without wishing to demean the motives of older users we hypothesize that they needed to assign some constructive purpose to tripping to justify their use of drugs which were then seen as more dangerous and powerful.

We anticipate that as the psychedelic-using population grows, recreational use will increase and, as with marihuana, will become less ritualized although not less controlled. We do not expect, however, that psychedelic-using rituals will ever approach the degree of flexibility and diversity of marihuana-using rituals. Quite probably psychedelic use will become more acceptable and social sanctions more available; but because of the high impact, long duration drug effect and the related tendency to keep psychedelic use infrequent there is both less need and less social opportunity to internalize social sanctions. Thus, there will remain a dependence on rituals (on external controls) which should limit the flexibility and diversity of psychedelic use.

Opiates

The larger culture condemns the illicit use of opiates more than any other drug. Popular mythology about the evils of the opiates and heroin, in particular, extends deep into the drug subculture itself. Many of the marihuana and psychedelic users in the DAC study do not recognize the possibility of controlled opiate use, even though they have identified and dispelled many of the larger culture's myths about their own drugs of choice.[20]

The controlled opiate users[21] in our study are painfully aware that they are seen as deviant. They tend to keep their use a closely guarded secret from everyone but their one or two dealers and other controlled opiate users. One of the researchers knew a woman he considered to be a reasonably close friend for several years, and although he had been previously involved in drug-related research, it was not until he became part of the DAC study that she felt free to "confess" that she had been a controlled heroin user all the while.

The relationship of controlled opiate users to addict/compulsive opiate users is as fraught with dangers and difficulties as it is necessary. One way controlled users can assert their normalcy is to spurn and condemn junkies, but they must rely on junkies to obtain opiates.[22] Addicts do not understand and are often threatened by controlled users' peculiar relation to opiates. So, on the one hand, controlled users get poor quality opiates at great cost from junkies ("You're always getting burned"), while on the other hand, they are repeatedly and seductively invited to become full-fledged members of the junkie subculture. The controlled user's constant dilemma is to become friendly enough with an addict to establish a reliable contact for quality opiates, but not so friendly that his refusals to fully participate in the addict's subculture insult the dealer who might then cut off the supply.

Beset on all sides, controlled users are bound together in small isolated groups that develop idiosyncratic, rigid rituals and social sanctions. These groups are fragile and drug-centered because it is difficult to find controlled users who are compatible as friends — the inverse of the situation with marihuana we described earlier.

Most of the rituals of controlled opiate users are indistinguishable from those of compulsive users. In both groups, people squabble over who gets off first, belts are used as **ties**, eye-droppers are used instead of syringes, booting is common, and **works** are cleaned but not boiled. The main reason for this ritual-sharing is that there is no highly visible, communicative population of controlled users from whom discrete rituals can evolve. Rituals are still being borrowed from the addict subculture — the only readily available source of expertise about the drug. There are also two other explanations for this phenomenon. First, while the life style of the addict is repugnant to most controlled users, they sometimes find the addict's bold outlaw stance attractive; partaking of the addict's ritual may be an expression of wistful identification. Second, several subjects were addicts before they became controlled users, and they have retained their former drug-using rituals (booting is probably the best example).

Several controlled users have added new elements to the addict ritual. One subject, for example, shifts the emphasis away from getting off by tacking on middle-class amenities — he plays the good host by serving wine and food to his user guests (this without any of the nausea which commonly accompanies opiate use) and all spend the evening together in conversation. Another user protects herself from a possible overdose by shooting a little of the drug, waiting to gauge its effect, and then shooting the remainder. By and large, however, controlled users' rituals are not well distinguished from those of compulsive users — especially in details of drug administration.

The social sanctions around controlled use are distinctive. Controlled users adhere to a variety of rules for opiates, most of which are summarized by the maxim: "Don't become dependent." They well appreciate that they can become addicted or compulsive users.

Ex-addict subjects have firm rules about frequency of use. One is a woman who has used heroin on an average of three to four times a month for over four years. Occasionally, when a break in her commitments to work and to her child permits, she goes on a using spree that lasts about a week. Even while on vacation, however, she will not use heroin more frequently than every other day. In general, subjects limit their opiate use far more than is needed to avoid addiction. One subject has confined his heroin use to weekends only for the past five years. One woman used heroin twice a month and on special occasions such as birthdays and New Year's, for a year and a half. Then, troubled by her tolerance to some of the drug's effects, she deliberately cut back use to only once a month. She ignored the fact that the variability in the potency of black market drugs could have accounted for her requiring the use of two **bags** instead of the usual one **bag** (on only two occasions) to obtain the same effects as when she used previously.

These and other examples indicate that many controlled users regard heroin as more rapidly addicting than is warranted, though they feel that it can be used moderately. This is understandable in view of the prevailing myths abcut heroin's power and the exposure controlled users have to addicts who have succumbed to the drug.

Controlled using subjects observe common sanctions against behaving like or becoming overly involved with junkies and compulsive users. Controlled users may chastize one another for manifesting irresponsible junkie-like behavior. Users who are unable to control the drug's effects may be chastized. A user of codeine-based cough syrup and of Doriden indicated that despite the somnolence induced by these drugs, people are expected to act responsibly: "One (cigarette) burn and you're thrown out." Being cheated by dealers is a fact of life, but a controlled user who cheats fellow users is punished by being called a junkie. Controlled users frown upon spending too much money on heroin because it suggests the junkie's lack of control: "Just 'cause I had the money don't necessary mean I would cop ... of course, I wouldn't steal to get the money to cop, there's no need for it 'cause I don't have a habit."

Shooting up like a junkie is O.K., but shooting up with junkies is not, because this symbolizes a loss of control. A couple who had regular access to opiates through the woman's addicted sister and brother-in-law stopped relying on them for opiates because of the social pressure to use the drug with them. They began borrowing a car and driving several miles to a copping site in another city where they knew they could obtain heroin from street dealers.

DISCUSSION AND NEW DIRECTIONS

Our findings show that, contrary to conventional wisdom, controlled use of illicit drugs is possible and is fostered by subcultural rituals and social sanctions that support controlled use and curtail drug abuse. We have also observed how the controlled use of alcohol is patterned by established, broad based rituals and social sanctions. These findings and observations strongly suggest that the evolution and widespread acceptance of social controls for illicit drugs, similar to those for alcohol, would provide a viable means of preventing drug abuse.

Ironically, the present attempt to eliminate all use of illicit drugs undermines users' ability to control them. Users receive no assistance from the larger culture for control. Instruction in how to use illicit drugs is now relegated to peer using groups which are, at best, an inadequate substitute for family-centered socialization. Association with controlled users is as much a matter of chance as it is of personal choice.[23] Because illicit drug use must be a covert activity, newcomers are not presented with an array of using groups from which to choose. Early in their using careers, many DAC subjects became involved with groups in which members were not well schooled in controlled use, or with groups in which compulsive use and risk-taking were the norms. In both cases subjects went through periods when drug use interfered with their ability to function and when they frequently experienced untoward drug effects such as bad trips. These individuals were later able to achieve controlled use, but many are not. To revoke personal commitments and realign oneself with new using companions is a difficult and again uncertain process.

The culture's active opposition to illicit drug use also alienates users from adult guidance. Asking adults for advice or approval even in a guarded way is risky, and raises difficult issues for parents and users alike. The deviant subcultures become more attractive because they insulate the user from the mainstream culture's disapproval and facilitate drug use.

Of course, the mainstream culture's opposition to illicit drug use is not wholly negative in its effects. Present legal and social sanctions do dissuade some people from taking these drugs and no doubt influence others to abandon their use, thereby preventing some unknown quantity of abuse. Unfortunately, it is not clear how many people would take these drugs if they were given an unobstructed choice about it, nor is it clear how many would go on to become abusers. What is clear is that the attempt to eliminate all use of these drugs contributes to their abuse by people who take them.

It seems safe to assume that no matter how massive the investments in law enforcement and education, neither the drugs themselves nor people's interest in taking them will be eliminated. There is every indication that illicit drug use will continue to rise as it has over the last decade. Given this prognosis and the failings and high social costs of our present restrictive social policy, it seems not only reasonable but necessary to place illicit drugs under social control so that their abuse can be minimized.

Ideally, social management of drug use affords advantages which prohibition does not. Drug use is normalized with other life activities and is transformed from a covert to an overt activity subject to the pressures of public scrutiny. Drug users regulate themselves and other users. Social learning in proper (controlled) drug use becomes available. Rituals and social sanctions provide freedom to pursue a recreational activity, albeit a complex and at times risky one, in an individualized way while discouraging detrimental drug-using behavior. Drug-taking loses its appeal as "forbidden fruit." Users who experience difficulties are more likely to seek assistance because they can do so without having to declare themselves deviant and morally bankrupt, and without the risk of punitive reprisals. The quality of drugs can be regulated and thus, untoward drug effects greatly reduced.

The chief difficulty in achieving social control over illicit drugs is that enormous changes would have to occur in both public attitude and social policy for effective controlling rituals and social sanctions to develop. Rituals and social sanctions cannot be supplied ready-made to drug users or potential users. We would, therefore, not recommend wholesale immediate legalization of marihuana, psychedelics, and the opiates precisely because too abrupt a shift in policy would leave many users without the elaborate social support needed to prevent abuse.

It is possible, however,

1. to alleviate major legal obstacles to their development, and
2. to provide more comprehensive and value-neutral information about licit and illicit drugs to the general population, making more user/non-user contact and discussion possible and, in turn, permitting further dissemination of controlling conventions.

Some steps could be taken now which would both strengthen the existing subcultural rituals and social sanctions and serve to demystify the power and danger of these drugs generally.

Certainly decriminalization of marihuana should be extended beyond those few states which have adopted it, and federal penalties for use should be dropped. Further research on the possible medical applications of marihuana and the psychedelics should be undertaken, and results sufficiently publicized so that their public image as "bad" drugs can be dissipated.[24] Heroin should be made available to physicians as a legitimate analgesic, and experimentation with heroin maintenance clinics for the treatment of addicts should also begin with careful control.

Drug education programs which are no more than disguised campaigns to eliminate use should be replaced with genuine efforts to provide users and non-users with some rudimentary pharmacological data and with detailed information about the consequences of various patterns of use. Doctors, teachers, counselors, and others who encounter drug users should be instructed in how to distinguish use from abuse — it simply makes no sense to alienate and undermine those segments of the population of drug-takers who stand against abuse.

These recommendations represent the first in a number of changes which would be required before illicit drugs could be made available under minimal legal restraints. We cannot detail here the entire sequence of such changes. In general, we recommend that social policy keep better pace with developments among drug users themselves than has so far been the case.

In closing, we suggest that the policy goals and changes we have outlined are part of a larger historical process by which drugs are gradually incorporated into a culture and by which use replaces abuse as a dominant using pattern. Turning again to alcohol as an example, in the seventeenth and eighteenth centuries 75 to 80 per cent of those who drank were alcoholics (Harrison, 1964). A few decades ago alcohol use was prohibited and the temperance movement pronounced it an evil and dangerous substance. Today 95 per cent of those who drink are controlled users. This figure might still be improved by further normalizing and not glorifying alcohol use, e.g., by banning advertising which relates alcohol use to sexual prowess.

In fact, illicit drugs are much further along in the process of becoming acceptable and controllable than the culture has been willing to acknowledge. If the incidence of untoward drug effects is an indication, we can see clear movement with respect to marihuana and the psychedelics. Becker (1963) notes that shortly after World War I the incidence of "panic reaction" to marihuana was higher than in the mid-1930's by which time marihuana use had increased in a number of groups. Today, such reactions are quite rare and are more typical of older (30+) users who have had no prior experience with marihuana. A few years ago the treatment of bad trips (resulting from use of psychedelics) accounted for as much as 20 to 35 per cent of hospital emergency admissions. Since that time psychedelic use has grown at a faster rate than the use of any other illicit drug (Drug Use in America, 1973), but the number of hospital admissions has dropped markedly. As of July 1974 the Massachusetts Mental Health Center did not know when they last had such an admission, but they were sure that it had been years rather than months ago (Grinspoon, 1974). The Haight Ashbury Free Medical Clinic, which furnishes emergency medical teams to rock concerts, reports (Smith, 1975) that at a recent concert attended by some 10,000 persons where psychedelics were openly distributed only two adverse reactions came to the attention of the medical team. In both cases, the patients were quickly quieted by talking with members of the team and sent home after fifteen to thirty minutes. A recent National Drug Abuse Council Survey Project shows that the majority of college and high school students who use drugs cannot be distinguished from many of those who do not and never have used drugs (Yankelovich, Skelly & White, 1975; Yankelovich, 1975).

These data suggest that the development of controlled using patterns for illicit drugs by substantial numbers of users is probably a recent occurrence. The legal system is not able to and probably should not reflect every shift in using patterns. But, if controlled using patterns stabilize, as our work indicates they have for marihuana and the psychedelics, and viable social sanctions which permit this use develop, then in time the laws should respond to the new social position of the illicit drug and the drug user. Obviously, it is difficult to develop rituals and social sanctions which are against the law; both the drug user and the public must tolerate a serious amount of ambiguity and anxiety. The user takes real risks by breaking the law (greater risks than are imposed by the chemistry of the drugs), and the public suffers the disruptions of laws which now punish more than they deter.

It does not seem likely that this situation will be rectified immediately. However, it is possible to monitor changing using patterns in order to determine how best to integrate these changes into the legal system. Until now there has been considerable resistance not only to legal changes but even to recognition of changing drug-using patterns. The study and dissemination of new information on how people develop successful drug-using patterns can proceed without neglecting the study of drug abuse when it occurs. Our work shows that controlled use of illicit drugs exists in this country and is the result of subcultural rituals and social sanctions.

NOTES

1. The material for this paper was gathered as part of a study of the social basis of drug abuse prevention funded by The Drug Abuse Council, Inc., 1828 L Street, N.W., Washington, D.C. The work of Richard C. Jacobson and Deborah Patt on that study was invaluable to this paper. Since July 1, 1976 research on controlled use has continued under National Institute on Drug Abuse Grant No. 1 R01 DA 01360-01A1.
2. "In more tribal cultures social sanctions are rarely institutionalized in a body of abstract law. Principles of rightness which underlie the activities are largely tacit. And they are not the subject of much explicit criticism, or even of very much reflective thought ... Legislation, though it may occur, is not the characteristic form of legal action" (Redfield, 1971).

3. Presumably drug use for religious purposes, such as the use of peyote by members of the Native American Church, would qualify as a ritual in the more classical sense.
4. Wilkinson (1970) reviews the relevant research in his Appendix A. Several references to Wilkinson follow as his work on alcohol closely parallels our own on the social determinants of controlled illicit drug use.
5. In many families the formal offer of a drink constitutes an important quasi-**rite de passage** from adolescence to adulthood.
6. Research has shown that in abstinent families where parallel conditions exist with respect to alcohol, the potential for children becoming alcoholics is greatly enhanced (Wilkinson, 1970, Appendix A).
7. The MEDLINE file contains 400,000 citations from 1,100 of the journals indexed for Index Medicus.
8. Indigenous data gatherers were trained in interviewing technique. All interviews were tape recorded, allowing research staff to monitor their work. Three of the data gatherers were recruited from within the sample — two women and one man — and proved extremely capable and reliable. They contributed the bulk of the data which were not gathered by the research staff.
9. Interviews have also been conducted with approximately 20 non-controlled drug users. Especially at the outset of the DAC study, potential subjects were referred to us who turned out, in fact, to be compulsive users. Interviews with these subjects provided valuable comparative data and were used as a basis to refine the interview schedule.
10. Further information on methods and other aspects of our findings are reported elsewhere: (1) R. C. Jacobson & N. E. Zinberg, 1975, The social basis of drug abuse prevention. Drug Abuse Coucil Special Studies Series, SS-5. Washington, D.C.: The Drug Abuse Council, Inc. (2) N. E. Zinberg, 1975, Addiction and ego function. The Psychoanalytic Study of the Child, 30:567-588. (3) N. E. Zinberg, R. C. Jacobson & W. M. Harding, 1975, Rituals and social sanctions as a basis of drug abuse prevention. The American Journal of Drug and Alcohol Abuse, 2:165-182. (4) N. E. Zinberg & R. C. Jacobson, 1976, The natural history of chipping. The American Journal of Psychiatry, 133:37-40.
11. While the influence of personality, family background, social class, availability of the drug, and other variables on drug use could be traced for individual subjects, no consistent relationship has been found between these factors and controlled use.
12. Weil & Zinberg (1968) found differences in ability to control the drug high among naive and experienced marihuana users in a controlled setting. Becker (1963) observed that users' appreciation and control of the drug high is learned; and that this learning allows the user to function adequately while under the influence of marihuana.
13. Users of the psychedelics and opiates were also able to control their highs but found it more difficult and usually limited use to protected settings.
14. Most Americans view marihuana as an illicit, "bad" drug, but as less "bad" than heroin, LSD, cocaine, etc.
15. In effect, new users recapitulate many of the elements of marihuana rituals of the early sixties in their preliminary use of the drug.
16. We found it more difficult to locate marihuana and psychedelic abusers than controlled users. This situation was reversed for the opiates.

17. Access to correctly labeled psychedelics is confined to a few knowledgeable, experienced, and well connected users. One user in the DAC sample was able to obtain psychedelics from a reputable source, and often had the drugs tested by a chemist before use.
18. There are substantive differences in the high states induced by the various psychedelics, which are beyond the scope of this article (Zinberg, 1974).
19. Psychedelics are chemically related to amphetamines. We are presuming here that these compulsive users are, in fact, using psychedelics and not wrongly labeled amphetamines.
20. Standing with the larger culture against opiate use may help marihuana and psychedelic users to view their own drug use as comparatively "good".
21. The preponderance of controlled opiate subjects were heroin users who used dilaudid, codeine, and other pharmaceutical opiates on an occasional basis. Only three subjects did not use heroin (see footnote following). Therefore, discussion will center on heroin use.
22. Three controlled users had regular access to opiates without going through a dealer: a physician who used morphine; a hemophiliac who could obtain pharmaceuticals from physicians under the pretense of relieving the pain of a hematoma; a user, whose drug of choice was codeine, who obtained cough syrup from a pharmacist willing to ignore existing legal regulations. These cases are described in some detail in Zinberg & Jacobson (1975).
23. This is less true for marihuana users than for psychedelic and opiate users.
24. We are not assuming that the results of this research will be uniformly positive. Whatever the results, by making these drugs the object of medical research the idea that no drug is inherently "good" or "bad", that any drug can be used in a variety of ways, would be advanced.

BIBLIOGRAPHY

Abrams, A., 1972, Accountability in drug education. Drug Abuse Council Monograph Series HS-1. Washington, D.C.: The Drug Abuse Council, Inc.

Becker, H. S., 1963, Outsiders: studies in the sociology of deviance. Glencoe, Illinois: Free Press of Glencoe.

Boston Sunday Globe. December 1, 1974, p.A-2.

Chein, I. et al., 1964, The road to H. New York: Basic Books.

Dobbs, W. H., 1971, Methadone treatment of heroin addicts. Journal of the American Medical Association, 218:1536-1541.

Freedman, Daniel X., M.D., 1971, Non-pharmacologic factors in drug dependence. Presented at a conference on the Non-Medical Use of Dependence Producing Drugs - Current Problems and Approaches. Geneva, Switzerland, October 20.

Gluckman, Max, 1972, Les rites de passage. In Max Gluckman (ed.), Essays on the ritual of social relations. Manchester: Manchester University Press, pp.1-52.

Goody, Jack, 1961, Religion and ritual: the definitional problem. The British Journal of Sociology, 7(June).

Grinspoon, Lester, 1974, personal communication.

Harrison, B., 1964, English drinking in the eighteenth century. New York/London: Oxford University Press.
Heller, M., 1972, The sources of drug abuse. Addiction Services Agency Report, June.
Kaplan, J., 1971, The new prohibition. New York: The World Publishing Company.
Keniston, K., 1968-69, Heads and seekers: drugs on campus, countercultures and American society. The American Scholar, 38.
Klausner, Samuel Z., 1964, Sacred and profane meanings of blood and alcohol. Journal of Social Psychology, 64:27-43.
LaBarré, Weston, 1938, The peyote cult. New Haven: Yale University Press.
Leach, Edmund R., 1968, Ritual. In David L. Stills (ed.), The International Encyclopedia of the Social Sciences, Vol.13:520-526.
Newmeyer, J., 1974, Five years after: drug use and exposure to heroin among the Haight Ashbury Free Medical Clinic clientele. Journal of Psychedelic drugs, 6:61-65.
New York Times, April 9, 1973.
Powell, D. H., 1973, Occasional heroin users: a pilot study. Archives of General Psychiatry, 28:586-594.
Redfield, Robert, 1953, The primitive world and its transformations. New York: Cornell University Press.
Scher, J., 1961, Group structure and narcotics addiction: notes for a natural history. International Journal of Group Psychotherapy, 11:81-93.
Scher, J., 1966, Patterns and profiles of addiction and drug abuse. Archives of General Psychiatry, 15:539-551.
Smith, David, 1975, personal communication.
Szasz, T., 1975, Ceremonial chemistry. New York: Anchor Press.
Weil, A., N. Zinberg, & J. Nelsen, 1968, Clinical and psychological effects of marihuana in man. Science, 162:1234-1242.
Wilkinson, R., 1970, The prevention of drinking problems. New York: Oxford University Press.
Yankelovich, D., 1975, How students control drugs. Psychology Today, Oct., 39-42.
Yankelovich, Skelly & White, 1975, Students and drugs: preventing drug abuse in the high schools and colleges. Drug Abuse Council Monograph Series, to be published.
Young, J., 1971, The drugtakers: the social meaning of drug use. London: MacGibbon & Kee, Ltd.
Zinberg, N., 1974, High states: a beginning study. Drug Abuse Council Special Study Series SS-3, Washington, D.C.: The Drug Abuse Council, Inc.
Zinberg, N.E. & R.C. Jacobson, 1976, The natural history of "chipping". The American Journal of Psychiatry, 133:37-40.

MODERN URBAN AMERICA

Turning from the international frame of reference in the previous section, the chapters in this third section deal with drug use among urban sub-cultural groups. The very fact of belonging to a set who shoot heroin, use methaqualone or partake of some other illicit drug sets these persons aside from mainstream American culture. It forms them into a sub-culture.

In two of these papers, however, we are also dealing with other contributing factors. Cleckner reports on Southern urban Blacks, while Sandoval discusses a gay bar crowd of Spanish Americans, many of whom were of Cuban extraction.

These papers reflect the growing trend among anthropologists to "get where the action is!" It reflects the increasing concern with ethnic, social, and behavioral problems and an attempt to improve these by research and understanding. But as important is the fact that all three of these authors represent a new breed of anthropologists who are being hired by Federal, State and special agency commissions because of their understanding of ethnic and cultural variability.

MICHAEL H. AGAR

7

INTO THAT WHOLE RITUAL THING: RITUALISTIC DRUG USE AMONG URBAN AMERICAN HEROIN ADDICTS[1]

The notion of 'ritual' has long been a topic of interest in anthropology, as well as in numerous other disciplines and in everyday language. While often discussed in connection with 'religion', the term is also used to more generally denote any highly predictable sequence of behavior. Within the drug field, for example, 'ritual' can refer to a frequent pattern of marijuana smoking wherein the users sit in a circle and pass a single cigarette from person to person.

Among urban American heroin addicts, 'ritual' is often used in everyday speech. For example, one addict might say to another, "yeah, he's into that whole ritual thing, man". In this usage, 'ritual' refers to one of several events that street addicts engage in on a day-to-day basis; namely, that of 'getting off', or administering the heroin intravenously. In this instance, folk and scientific usage comfortably blend, for getting off is also the one event in the addict repertoire that best conforms to the more formal notion of 'ritual' as a predictable sequence of behavior.

For the purposes of this symposium then, getting off will serve as an excellent example of a drug related ritual event. First, I hope to describe the general outlines of the event. Next, I will formally contrast it with other addict events in an attempt to characterize what it is that makes getting off more 'ritual' than the others. Finally, some ritual implications of the recent shift to methadone as a street narcotic in New York will be presented.

The content of the paper is based on two years' fieldwork with heroin addicts at the NIMH Clinical Research Center in Lexington, Ky., brief associations with the Waikiki Drug Clinic in Honolulu, Hi., and the Haight Ashbury Free Clinic in San Francisco, and, most recently, a year and a half with the Drug Abuse Control Commission in New York City. First of all, the getting-off event will be briefly described, adapting for present purposes descriptive materials published as part of a previous study (Agar, 1973).[2]

The ritual event considered here, as already noted, is labeled getting off ('taking off', 'shooting up', 'fixing'). To get-off, the 'junkie' (addict) must have the necessary paraphernalia. Among these are the set of 'works' ('outfit'). There is some disagreement as to just what is included in the works. All agree that it must include a hypodermic needle or 'spike' ('pin') and either a 'gun' (hypodermic syringe) or a 'dropper' and 'bulb'.

The spike size is usually 26 gauge. The dropper may be an ordinary glass eyedropper or a dropper from Murine, a commercially available eye relaxant. The Murine dropper may be preferred because the 26 spike fits snugly over the end. Because of its relatively small size, though, many will not use it. Since most junkies have some trouble getting a 'hit' (needle into the vein) they want to use all the heroin they intend to in one fix. This often requires more volume than the Murine dropper can provide. Thus, many will prefer the larger dropper.

If the ordinary dropper is used, some thread is wrapped around the tip to ensure a snug fit with the spike. Another type of 'collar' is the 'Gee', a strip of paper torn from a $1.00 bill and similarly wrapped. Rather than using the standard rubber bulb, the bulb from a baby pacifier is removed and fastened to the top of the dropper. The larger bulb is preferred, since it increases both the pressure that can be used to 'draw up' the heroin and the total volume of the works.

Although some junkies use a gun, most will not for two reasons. First, the injection must usually be administered with one hand, since the other usually holds the 'tie' (glossed later). The dropper is much easier to handle with one hand than is the gun, especially if one wants to 'boot' (glossed later). Second, most addicts must probe to get a hit. With the dropper, a hit is indicated when the pressure in the vein forces some blood into the works. This is called a 'flag' ('register'). With a gun, the junkie must pull back on the plunger to test for a hit. This often jerks the spike out of the vein. Thus, unless a junkie has good 'ropes' (veins) that are easy to hit, he will prefer the dropper.

Three other items are included in the outfit by some addicts: a small piece of wire is useful to clean the spike if it becomes clogged. A razor blade may be included to divide a quantity into equal piles. Cotton may be carried and used to filter the dissolved heroin into the works. The cotton catches the undissolved material and minimizes the possibility of clogging the spike. Others may use part of a cigarette filter or a piece of Kleenex for this filtering process. In addition, the junkie will need a 'cooker'. This can be any small, nonflammable receptacle in which the heroin can be dissolved. Two frequently used cookers are a spoon with the handle bent to prevent spillage or a bottle cap with a hairpin handle attached.

Finally, the addict needs a 'tie', unless he has exceptionally good ropes. A tie is any flexible material that can be applied like a tourniquet to force the veins to stand out. This facilitates getting a hit. Some examples of ties are a belt, a nylon stocking, or a thin piece of cord. If a belt is used for an injection into the arm, for example, the belt is wrapped around the mid-upper arm and chinched tightly. The junkie, who is sitting, then leans forward on the elbow of the belted arm with the belt running under the elbow to the left hand. Held tightly, the veins bulge and the right arm is free to use the works. By releasing his grip, the tension on the belt is relaxed.

This is only one example. There are different ways of using different ties. Furthermore, the arm is not the only area for injection. Any vein can be used, for example, in the hands, legs, or feet. Some hit in the neck, and there are rarer cases who use a vein in the tongue. Other veins are usually used sooner or later, since constant use of one vein usually results in venal collapse. Furthermore unsterile needles or accidental subcutaneous injection may cause abscesses, also necessitating a move to another spot.

Given the necessary equipment, the addict must know how to use it. First, he must 'crack' (open) the bags and place the heroin in the cooker. The works are filled with water, and the water is squirted into the cooker. He then heats the mixture over a burning match or candle to dissolve the heroin. The cooker is set on some surface and the cotton (or other filter) is dropped into the mixture. The tip of the needle is placed in the cotton and the mixture is drawn into the works. The tie is applied, and the addict probes for a vein until the flag comes up. Usually, he then loosens the tie to ensure that the needle does not slip out. He can then shoot-up the entire fix, or he can 'boot' ('jack', 'milk').

If he boots, he shoots in a fraction of the fix, then releases the bulb, thus drawing in blood. He then reapplies pressure to the bulb, shooting the blood-heroin mixture ('gravy'). He can continue to boot until the works are empty of heroin. He may do this for one or two reasons. First, he may want to test the quality by fixing a bit at a time. Second, he may want the multiple 'rush' (glossed later) that comes with booting. The disadvantages of booting include the possibility of clogging the needle as the mixture cools. Finally, one can boot, obviously only when one has adequate time to do so.

As the drug first enters the body, the junkie experiences the 'rush' or 'flash' (initial physiological effects). The rush, as well as the other physical effects discussed later, are usually said to be impossible to describe to a non-addict. The rush is sometimes compared to a 'driving force' or to an orgasm. Two important implications of the rush should be noted. First, popular knowledge has it that heroin use has no effect on a junkie after his tolerance builds up. While this is true for some effects (to be discussed), it is not for the rush. Unless he has purchased a 'blank' (bag of fake heroin), the junkie experiences the rush no matter how addicted he is. A second implication is that intravenous injection is preferred because it is a means to a quicker, better rush. 'Snorting' (sniffing heroin like snuff) and 'skin-popping' (subcutaneous injection) are possible techniques of administration, as is simple ingestion. But none of these produce a rush as rapid or powerful as intravenous injection.

A second kind of effect is the 'high', described as a feeling of general well-being. This effect (and the 'nod', discussed later) decreases with increased tolerance. That is, as the junkie becomes more addicted and acquires a higher tolerance to heroin, it takes increasing amounts to make him high. The high is longer lasting than the rush, though the length of time varies with the tolerance of the junkie and the dosage.

A third effect is the 'nod', usually described as a state of unawareness, a kind of chemical limbo. Nods can vary from 'light' to 'heavy'. A light nod produces such effects as slightly dropping eyelids and jaw, whereas a heavy nod is a state of complete unconsciousness. The nod is less frequent than the high, since a higher dose of heroin relative to the junkie's tolerance is necessary to bring it about.

A fourth effect is the feeling of being 'straight'. A junkie is straight when he is not sick. If the fix removes whatever withdrawal symptoms the junkie is experiencing, then it gets him straight. Four to six hours after the injection, symptoms of withdrawal begin to appear (runny rose, watery eyes, chills, among others) and continually worsen until another shot is administered.

Unless he has purchased a blank the junkie will get a rush and get straight. Depending on the amount in his fix and the quality (i.e. 'garbage' (poor quality), 'decent' (normal), or 'dynamite' (high quality), he may get high and possibly nod. If the dose is too high or the quality too good, he may 'O.D.' (overdose) and die.

THE EVENT AS RITUAL

As mentioned earlier, getting off is only one of a number of events engaged in by heroin addicts. One 'hustles' to get money and 'cops' to obtain heroin. One worries about being 'busted' by the police, 'hangs out' with other addicts, and 'parties' with addict and non-addict friends. Why is getting off more of a ritual than these other events? A first answer is simply that getting off is more predictable. If a member of the subculture knows that an addict is about to get off, he can make more statements about what that addict will do than he can for any of the other events. Or can he?

Getting-off is not all that predictable. No particular style of dress is prescribed; any of a number of possible physical settings can be used, ranging from a hallway to a public restroom to a comfortable apartment. Although there is traditionally an argument over who gets off first, much verbal exchange can focus on a variety of topics related to other events as well. In fact, in some ways copping is more predictable — it almost always occurs on or near certain street corners or in a specific dealer's residence, and the verbal exchange is more narrowly focused on the economics of the purchase. Perhaps copping should be called a ritual.

There is one part of getting off where high predictability does differentiate it from the other events. The sequence of psychomotor activities in the event is most rigidly prescribed — the preparation of the heroin, its transfer to the works, and its injection. Even here, though, getting off may not be unique. For example, certain hustles have predictable sequences of psychomotor activity, such as lock-picking. On the other hand, one can argue that knowing a person is about to hustle does not generally allow many statements to be made concerning that person's subsequent motor behavior. The hustle used, and the specific form of that hustle, depend on several contingencies that cannot be anticipated. An addict often does not know what opportunities will be available in the streets until he arrives there. Further, he may not be sure what type of person he will deal with, and different types, depending on the hustle, may require different strategies. In contrast, getting off is likely to take quite similar psychomotor form whatever the context.

This discussion suggests a first unique feature of the ritual event:

For an event to be a ritual event, it must prescribe a sequence of psychomotor acts.

Of course, other aspects of a ritual event performance may also be prescribed — dress, physical setting, verbal content, and so on. But none of these are either necessary or sufficient to constitute a ritual, or so it is hypothesized here.

In spite of the counterarguments, though, an addict going through the carefully prescribed motions of the lock-picking hustle easily fits the definition. What might differentiate the lock-picking hustle from getting off? As a first answer, perhaps one sequence is invested with a 'special meaning' by its participant, while the other is not. Shortly, the loose notion of 'special meaning' is dealt with, but for now it is used intuitively

The notion of special meaning is appealing in the example used here. For a heroin addict, the getting off sequence clearly has a special meaning. It represents the core of his life as an addict, the eventual goal around which much of his effort is focused. This leads to a second feature of a ritual event:

For an event to be a ritual event, the prescribed psychomotor sequence must be invested with a special meaning for the person performing that sequence.

The problem, of course, lies in determining when a psychomotor sequence has 'special meaning' for a person. Clearly this is a difficult area, though I think an unavoidable one. Several indicators might be useful. First, of course, is the expressed attitude towards the event when it is spoken of out of context. Without going into detail, I think anyone who discussed getting-off with a group of addicts, in a context where the addicts feel that no moralistic judgments are being made, will hear much positive affect, indeed, the correct word may be 'reverence', expressed toward this event.

A second indicator might be the use of something related to the psychomotor sequence for group emblematic purposes. Again, getting off fits this description. The works, for example, are a frequent item of grafitti in the residential and day rooms of treatment centers. The argument here, then, is that some integral part of the total sequence, in this case the works, becomes a shorthand emblem for one's identity because of the special meaning attached to the sequence as a whole.

A third indicator of special meaning might be reflected in a tendency of many group members to become obsessed with the sequence. In the case of getting off, the obsession appears in at least two ways. First of all, some significant, though unknown, percentage of the addict population is known as 'needle freaks'. In the argot, this term refers to those to whom the process of getting off is equally or more important than the use of heroin itself. Second, several addicts on the streets during the recent period of low-quality heroin said they would sometimes buy a bag anyway, even though they knew it was garbage, just to "do that whole thing again, man, you know". These examples, then suggest that special meaning may be indicated by either obsessive performance of the ritual, or continued performance, at least for a time, even after the rationale is no longer present.

These three attempts to find empirical indicators of 'special meaning' are tentative at best. The notion is obviously a difficult one, yet on an intuitive basis it seems crucial. 'Ritual' is not intended to refer to all prescribed sequences of behavior, I don't think. However, adoption of a criterion of 'special meaning' dangerously blurs the distinction between 'sacred' and 'secular', a distinction near to the heart of this symposium. How does 'special meaning' fit into this traditional distinction? If special meaning is attached to getting off by the addict, is the ritual sacred or secular for the addict?

'Sacred' and 'secular' are concepts used by social scientists, and perhaps one way of resolving the problem is to assert that whatever has special meaning for a group member counts as an instance of sacred ritual. The problem for the social scientist, then, is to be sure that he has correctly understood the meaning of some event for the group members, for if he doesn't, he will incorrectly translate the event into the dichotomy of sacred versus secular.

The problem is compounded when the social scientist-translator is himself a member of a society with strong negative attitudes towards heroin use. To allow himself to think of any aspect of a heroin addict's life as 'sacred' would contradict his own firmly rooted biases about the nature of sacred ritual, the heroin addict, and the total incompatibility between the two. Thus, while I thought about and wrote this paper, I decided that getting off was a secular ritual in my eyes, but a sacred one in the eyes of the heroin addict — or at least in the eyes of some heroin addicts, and therein lies a second complication.

Not all addicts are as attached to getting off as others. As already noted, some are obsessed with the ritual — needle freaks. Others, however, view getting off as strictly an instrumental act. They have no particular concern with the order or the manner in which things are done; they simply want to get the heroin into their systems. Most addicts, of those I have known, place the event somewhere in between, in an area fitting the notion of ritual as a psychomotor sequence with special meaning for them.

This only documents the obvious fact that for any ritual in some group, there will be intragroup variation in the extent to which the event is perceived as a ritual. Two examples might be useful here. In the first, note that an area traditionally thought of as religiously ritualistic — a Catholic mass — is sometimes attended solely for pragmatic reasons; no special meaning attaches to it. In contrast, there are bowlers for whom bowling is truly a ritual event. They talk about it frequently and positively in other contexts; their homes are adorned with trophies and their jackets bear patches from teams and tournaments, and so on. Thus, any event with a prescribed psychomotor sequence is potentially a ritual for at least some of its participants, and, very likely, no event is a ritual for all the individuals who participate in it.

METHADONE ISN'T FOR SHOOTING

The injection of heroin, useful as it might be to exemplify a drug-related ritual, may become a historical case rather than a contemporary example. There are recent indications that a crackdown on heroin, together with an explosion in the number of available methadone program slots, have dramatically altered the street narcotics scene in New York City. Although it is too early to predict the final outcome of these shifts, recent research suggests that methadone, as of summer, 1974, was becoming the most frequently used narcotic in the streets.

Along with the shift from heroin to methadone, a shift in the method of administering the drug has also occurred. Heroin, after some initial experimentation, is almost always injected intravenously. Methadone, on the other hand, is almost always taken orally, usually mixed with some flavored drink. Since the previous two sections describe the intravenous use of heroin as a ritual for the addict why isn't methadone used intravenously as it becomes integrated into the streets?

At least two types of answers are given by methadone users. The first simply asserts that methadone, as administered by the clinic, is not injectable. For example, an addict might state that the mixture cannot be cooked down in order to obtain injectable methadone. Others will note that no method of preparation will produce a liquid fluid enough to pass through the hypodermic needle.

It is difficult to evaluate such explanations, especially since the form of methadone administered by different clinics sometimes changes. However, one professional study (Jaffe et al., 1971) suggests that the tablet or disquette form is injectable, while liquid forms may or may not be injectable, depending on the additive liquid. Further, there is some limited data indicating that methadone in both liquid and tablet forms can be injected (Agar & Stephens, 1975; Agar, 1975). However, most data overhwlemingly indicate that methadone is for drinking, not for shooting.

The second type of explanation is simply the classic one of cultural inertia – i.e. "that's just the way we do things". Although methadone, known primarily by another name (dolophine), has been around a long time, its widespread, high frequency street use as an euphorigenic agent has only occurred since its recent popularity as a treatment modality. Because the clinic was, and continues to be, the primary source of supply, methadone is received in a context where the oral route of administration is the 'natural' way of using it. Methadone may be used orally primarily because that is the 'traditional' way that an addict conceives of methadone use.

In short, methadone is to drink, either because its chemical composition prohibits injection, or because an addict's cultural syntax is likely to link methadone with drinking, heroin with shooting. One of the unintended consequences of the methadone maintenance treatment modality, then, may have been the undermining of the key ritual in the street heroin addict's daily activities. By changing narcotics use into the simple act of drinking from a clinic cup or bottle, the event changes in several ways. Gone are the elaborate preparation, and the need for sharing works because illegality made it risky to carry them. No longer is a place needed, since use now takes almost no time. It can be done in seconds anywhere – on a streetcorner, in a bus, and so on. The carefully prescribed sequence of activities outlined in the earlier descriptive section of this paper was simply done away with by the shift to an oral route of administration.

Now, as an anthropologist, it would be interesting to relate this rather important aspect of change in the ritual domain to many other things, such as the efficacy of methadone maintenance, the changes in the current drug scene, and so on. As in the classic case of the Yir Yiront in Australia (1952), discussed in the well known article 'Steel axes for Stone-Age Australians', we could show that a simple technological change – the shift from works to the clinic cup as the route of administration – had effects that pervaded different areas of a culture and eventually destroyed it.

However, the explanation is undoubtedly not that simple. In fact, the case of methadone may enlighten us generally on the oversimplified attribution of causation to ritual change. In addition to altering the getting off event, methadone affected the street addict subculture in many other ways. For example, the widespread availability of an inexpensive, powerful narcotic, easily available in small amounts from a large number of people, dramatically altered the economic structure of the street market. In the earlier heroin market, street narcotics were highly diluted, expensive, and controlled in large amounts by a relatively small number of high level dealers. Perhaps this 'economic' change is a better explanation of methadone's impact. Consider yet another alternative. Methadone is the first recent treatment modality to simplify the transition from 'addict' to 'patient'. In drug free approaches, the addict must withdraw from the narcotic before commencing treatment, while methadone allows him to continue narcotics use while treatment is ongoing. Perhaps this weaker, more flexible boundary line between 'addict' and 'patient' status is a more plausible explanation for methadone's impact on the old heroin addict street subculture of the late 1960's and early 1970's. Other examples could also be offered as explanations, but these two serve to make the point.

So, how does one evaluate the role of ritual change in the recent shift from shooting heroin to drinking methadone? Is the change a cause, an effect, or an unexpected correlate of methadone's impact on the heroin-using street scene? Or perhaps the 'change' is only illusory, at least in the area of ritual, and soon methadone-centered groups will also develop ritual events with methadone as an integral part. Finally, as noted earlier, heroin may eventually return in both quantity and quality, revealing methadone use as a temporary adaptation of hard times.

At present, the developmental trajectory of the street narcotics scene is far from clear. Press reports and anecdotal data contain contradictory predictions, and to some extent developments will depend on policy decisions that have not yet been made. In spite of this current ambiguity, though, I thought it worthwhile to point out that a change in narcotic availability from heroin to methadone simultaneously:

1. had an impact on the economic structure of the heroin scene and proved efficacious as a treatment modality with a higher percentage of heroin addicts; and

2. dramatically altered a key ritual event in the heroin addict subculture.

This cooccurrence could hardly be accidental.

CONCLUSION

By examining the structure of a 'folk-ritual' event among urban American heroin addicts, a tentative formal definition of 'ritual' has emerged.

The definition, stressing the two features of a prescribed psychomotor sequence together with a perceived special meaning for that sequence, is certainly in need of broader application in other cultural settings as a further test of its adequacy. This definition does unite a variety of activities holding 'special meaning' for their participants; intuitively, I think they belong together as practitioners of 'ritual'.

Of course, there are many other anthropologically interesting modes of viewing getting off among heroin addicts as a ritual event. One could, for example, take a cue from Radcliffe-Brown and examine the event for an engendered sense of social solidarity among addicts. The application would not be a simple one, since many addicts, at least some of the time, inject the heroin alone, either because of time pressure or simple preference. On the other hand, there are occasions of social use where a sense of 'we' versus 'they' is definitely articulated. But then, to further complicate matters, such a message is also explicit in other social settings where heroin is not used, such as when hanging out in public places, or when talking in non-street settings like prisons and treatment centers. In short, some occasions of getting off may engender social solidarity among addicts, but this solidarity may also result from other, non-ritual settings as well.

Another classic master, Malinowski, would be equally difficult to apply in a straight-forward manner. Malinowski's notion of ritual as an uncertainty reducer would take on a new twist here, since the ritual involves the use of a pharmacologically defined uncertainty reducing drug. In fact, it would be difficult to use heroin at all in a non-ritualistic sense, using Malinowski's perspective. Further there are other complications. For example there is traditionally an argument over 'who gets off first' in social settings with limited works, an argument that can become quite heated. True enough, the argument is always resolved, but from one perspective it is an 'uncertainty producer' within the ritual. Because of the illegal nature of the event, anxieties are also voiced over an immanent bust that might occur before a person has a chance to get off. In short, uncertainty reduction in this event is a complicated issue, partly because the effect may be produced more by the chemical than by the behavior, and partly because the setting and rules of the event itself contain some anxiety producing aspects.

Those interested in the biologically based, culturally interpreted symbols of Turner, to take another example, would find some hints in the psychiatric literature. Psychiatrists sometimes discuss the sexual-metaphorical basis of the act of injection. For example, booting is also called 'jacking off' by some addicts; the rush is often compared to an orgasm: a line in a song by the Velvet Underground, sometimes quoted by addicts, goes, "heroin, it's my life and it's my wife". Although a symbolic interpretation of the actual act of fixing may be partially relevant, there are other notions surrounding the act, including some very pragmatic ones. For example, many addicts see shooting up not as a ritually symbolic act, but rather as the quickest method to the most powerful rush, much preferred over snorting, for example. Again, a symbolic metaphor provides a partial fit with this particular ritual event, while neglecting other crucial aspects.

As a final example of other applicable anthropological modes of viewing ritual, consider a Levi-Straussian notion of ritual as exchange. In social settings, where getting off occurs, there is often much sharing of works and heroin. Usually, there will not be enough works to go around, as mentioned above. An examination of the rules of sharing might lead into a deeper understanding of social relationships in the street addict culture. As another example, often one or two destitute members of the group will ask for a 'taste' from one of the group members who is flush. Narrowing in on such exchange-based interactions that are part of the ritual event would be useful. Again, the perspective would be a limited one, but an exchange focus suggests several interesting questions not treated here.

Rather than taking some established statements about ritual and testing the goodness of fit of 'getting off', the approach here has been more of an inductive one. First, an event was selected that both group members and social scientists would be likely to characterize as ritual. Then, the event was compared with others to try and efine the characteristics of the ritual event that might serve to differentiate it. The street addict culture is an excellent choice for this exercise, since it is probably more narrowly focussed on a smaller number of key events than many of the groups usually studied by ethnographers.

So, although many perspectives on ritual could be applied here, this is only a paper, not a volume. However, by focusing on the internal organization of one ritual event, and by contrasting this event with others, I have tried to explicate at least two principles by which the 'sacred' is sorted from the 'profane' in our own and in other societies.

NOTES

1. Time for the preparation and presentation of this paper was provided by the New York State Drug Abuse Control Commission, whose support is gratefully acknowledged.
2. Material in the section entitled 'Getting off' is adapted from material published elsewhere in 'Ripping and running' by Academic Press. See bibliography for full citation.

BIBLIOGRAPHY

Agar, Michael, 1973, Ripping and running. New York: Academic Press.
Agar, Michael, 1975, Going through the changes. Human Organization, in press.

Agar, Michael & Richard Stephens, 1975, Methadone in the streets: the addict's view. Psychiatry, in press.

Jaffe, H. H., et al., 1971, Methadone disks; injectable — noninjectable tablets. Archives of General Psychiatry, 25:525-526.

Sharp, Lauriston, 1952, Steel axes for Stone-Age Australians. Human Organization, 2:17-22.

PATRICIA J. CLECKNER

8

COGNITIVE AND RITUAL ASPECTS OF DRUG USE AMONG YOUNG BLACK URBAN MALES[1]

"IF YOU PLAY, YOU GOT TO PAY"

A great deal has been written concerning young Black men in the street including such ethnographic classics as 'Tally's Corner' (Liebow, 1967), 'Soulside' (Hannerz, 1969), and 'The Vice Lords' (Keiser, 1969). Their involvement in drugs has been extensively researched, and perhaps overresearched. Like college sophomores, they are a group which is accessible through institutional settings, often literally being captive subjects. They are considered to be the modal drug user group. In fact, they are probably not. Rather they are the modal group of identifiable users. There are several reasons for this. Institutionalization is an integral part of all street life. For these men the street and the institution, be it prison, jail or a rehabilitative program, are two co-existing and opposing settings in which to function. Life is a contrapuntal interaction between these two styles of being. Social agencies such as the court system and social workers differentially channelize Black male clients in these directions. Street men are institution oriented as individuals. Most significantly, they have a peculiar vulnerability to failure in using drugs: first, because of the difficulty in supporting heavy drug use due to limited economic resources and hence become involved in crime, and second, because of the standards which they set for themselves in terms of status and lifestyles. This chapter will explore this second issue in order to articulate the nature of this vulnerability.

In seeking explanations of drug use in the ghetto environment it seems to me that most researchers have overlooked some very obvious dynamics. As the men themselves would say, "They've shot past the money". The key to these dynamics lies in the ritual and symbolic content of the drug scene in the consciousness of the users themselves. Drugs have a meaning and serve a function which is intimately and profoundly linked to the activities and metaphors of street life. They serve a pseudotherapeutic purpose in relieving its unique anxieties and creating states of consciousness which articulate productively with the social reality. Perhaps the tragedy of the drug experience in the minds of these men is not that it leads to a life of crime, but that it does not live up to its promise. Its therapeutic and positive potential are deceptively seductive to these men. Drugs fit well into the values of the street: experience, cool, challenge and conspicuous consumption, but ultimately they fail. Heroin especially exemplifies this phenomenon:

> That first time is hellified and after that you can shoot and shoot and shoot and never get that high. You get high but it's never the same. That's the irony of it really because you're hooked by looking for that high.

One name of heroin is white lady. This term originated:

> Due to some sort of type of stigma that went along with creative Black people in the forties and the thing was the White lady was sweet but she was also dangerous and not loyal when she was with a Black man. She was always running out on you. She would get you in trouble and she couldn't be found when you needed her most.

The basis of these dynamics is not strictly social. Drugs are primarily a subjective experience. This is a simple but important point. To understand the relationship of drugs to the street, of their effect on the consciousness and action, it is necessary to understand the various highs which the users themselves experience. Michael Harner makes the point in discussing anthropologists' investigation of hallucinogens:

> Undoubtedly one of the major reasons that anthropologists for so long underestimated the importance of hallucinogenic substances in shamanism and religious experience was that very few had partaken themselves of the native psychotropic materials (other than peyote) or had undergone the resulting subjective experiences so critical, perhaps paradoxically, to an empirical understanding of their meaning to the peoples they studied (1973, Preface).

While it is debatable whether or not the practicing anthropologist should experiment with heroin, it is clear that careful attention should be given to the centrality of the personal experience of getting high. The users themselves indicate this in the way they talk and think about drugs. Without understanding the high, it is an enigma why anyone would use heroin. Stephens & McBride (1976:5) report that 42 per cent of addicts questioned on their attitudes prior to use reported that they didn't even like addicts. The fact of addiction is well known, yet most young men feel "it will never happen to them". This is some indication of the seductiveness of the drug. The basic appeal of 'feeling good' from drugs, supported by a variety of social, cognitive and ritual attitudes associated with the street code far outweighs the disadvantages.

THE SCENE

The data for this paper were collected in interview and participant-observation situations among Southern urban Blacks most of whom grew up together and participated in the same peer group activities. The area in which they grew up was the largest concentration of Blacks in the city, a ghetto, more like Watts than Harlem, resembling, in many ways, a rundown suburb. The entire area is open and sunny, a mixture of two or three storey projects and single family dwellings. Some of the streets resemble country roads.

The open-space nature of the area adds a unique flavor to illegal activities. In a close city like New York it is much easier to engage in illegal activities without detection. A shooting gallery located in an apartment building has a certain degree of anonymity since traffic in and out is almost indistinguishable from other traffic in the building. In this ghetto, however, traffic in and out of a single family dwelling is highly visible. One finds fewer shooting galleries here and the 'dope houses' which do exist usually have considerable ties with enforcement agencies or are short lived. Selling of drugs is more open often because it is safer to deal in the street where one can be on the lookout for uniformed police than in a closed building where one may be unaware of their approach.

Entertainment and drug activity often focused in local bars called clubs or juke joints. These places generally have a few tables and chairs, a pool table, a juke box, and only sell beer and sweet wine. In some of these clubs, 'reefer' or marijuana is smoked openly. Here men gather to relax and do business. Drug dealing is high competitive and is done outside the bar in order to spot both uniformed police and customers. Stashes are scattered throughout the street and rarely kept on the person. The presence of undercover police is unlikely in the area since the quantity of dealing is actually small and arrests are not worth 'blowing a cover'. In addition, most people in these places know each other and strangers are treated with caution.

The men I interviewed were in their early twenties. Most were second generation urbanites, their parents having migrated from nearby Southern states. All had experienced the pains of integration midway in their school career. Many of their life difficulties seem to be attributable to economic deprivation or to their parents' inability to equip them adequately for the urban environment, themselves having recently come to the city. These men relied heavily on their peers for education and preparation for life. Most of those I interviewed belonged to the same clique or gang, associating with many of the same people from the age of 10 or 11. In this respect their relationships were highly stable. During adolescence the peers in question were preoccupied with fighting and athletics with a highly competitive emphasis. Status was generally determined by who could fight the best or hit the hardest.

Most of the activities were oriented toward aggression or gang fighting. It had been suggested that these groups provide a way of relieving frustration and structuring aggressive impulses. That is, it is style or cultural pattern through which aggression is expressed Second, there is the psychological factor of group acceptance and reçognition. Third, it is possible that such groups serve as preparation for survival in the ghetto environment where physical prowess and mental quickness are absolute necessities. This includes self defense and knowledge of how to obtain money and goods in the absence of access ot elgitimate resources. Much fighting and 'ripping off' has a playful quality. These activities are engaged in for their own sake. For example, several of the group members participated in breaking into a school. Once they got inside they entered the kitchen and took the school's supply of butter and fried all the chicken they could find. The entire school was served cold chicken in subsequent days.

This playful attitude as well as close friendships continued into adulthood. Men enjoy practicing their conning skills on one another. A man will surreptitiously take a dollar or a half ounce of reefer (or anything else small) from another. This is not considered serious,

> Usually when you know people for so long like X yesterday. (The person in question had taken his reefer.) My thing was not gettin' too pissed off about it because they just left and took the reefer while I couldn't go with them.
>
> But yesterday I walked to X and I didn't have no bus money. I say, "give me $2.00" and he didn't think to hesitate to give the $2.00 and we joked about the reefer. It wasn't to rip me off but out of playfulness because everything goes. The least person I'd expect to rip-off my stereo would be him.
>
> They wasn't stealin' they was taken, not rippin' off but bull shittin'. It was in a jokin' attitude and it goes into a personal thing. As far as friendship it's what you could label horseplay.
>
> Most up on (the avenue) were into horseplay punchin' on each other and kicking and using BB guns.
>
> I wasn't pissed off because I knew it wasn't with somebody I could argue about. Its like if you loan somebody somethin' and forget to pay it back.

This behavior developed from and has continued after the gang. Perhaps it is a way of keeping people sharp and was a way of learning how to rip off in the first place. In a sense, it is the behavioral correlate of 'sounding' or 'ritual insults'.

It was against the back drop of this gang that people started using drugs as part of an informal peer education. Most members did not start using while the gang was in existence and not all of them use drugs now. Those members who are now users, ex-users or dealers have generally tried a large variety of drugs, including marijuana, barbiturates, amphetamines, heroin, cocaine, miscellaneous prescription preparations and occasionally psychedelics. In this respect, they are similar to young users in other cities throughout the United States in recent years.

GAMES AND IMAGES

The ideological and ethical basis for action among the men whom I interviewed conforms in large part with similar street codes which have been reported in the past (Finestone, 1962; Hudson, 1972). As such it is a variant of some larger value patterns in America, of what one informant called 'the American dream'. Status, competition and material display are emphasized but in a unique pattern which articulates Black values, perspectives and life situations. Status in the eyes of one's peers plays a dominant role. Considerable street time is taken up assessing others' statuses and reassessing and maintaining ones own. Status, along with actual material gain are the rewards in a sense of playing the street game. Money is a key focal point and primary goal for most street men and it is a major determinant of status. "The main thing about the game is makin' money and seeing who can look the slickest." This preoccupation is understandable in an environment of poverty and limited resources. It is a commodity in high demand and most of these men pay dearly for it.

Street ethics are dominated by a peculiar metaphor which profoundly reflects the mental orientation of these men. It consists of two related elements, the game and the play. It implies the importance of illusion and manipulation of a mental attitude which addresses the world as something other than what it seems, either as a kind of limited context such as a game which is played or as a theatrical event dominated by the creation of an illusion. These two ideas come together to form an image of what a man should be, he should be cool, detached and never caught up in the drama of which he is a part. He must be always cautious, always aware of possibilities and dangers at the same time appearing to be enjoying what he is doing. He must control his emotions and use the emotions of others to gain control over them.

The street is sometimes referred to as 'the game'. One informant described his early experiences as "I was still learning the game". Within the overall street game are more specific strategies, or 'hustles': the 'stuff game' or drug dealing, 'the pimp game' and the 'numbers game' or gambling. One informant defined it thus:

> Game to me is running stuff to different people, different ways of getting what you want. Game is hustling, conning or whatever.

Hassles with police are sometimes referred to as 'playing cops and robbers'. A man may tell a woman who leaves him that she has 'won the game this time'. People solicit favors or help by asking people to 'give them a play'. A con is called 'running game'. A person who overestimates the potential of a situation will 'play past the money'. The ultimate hustler who is winning the game is the 'player'. These are not mere linguistic accidents but reflect an entire mental approach which is dominated by strategies. Street men spend a great deal of time plotting approaches and looking for plays. When a young hustler 'gets open to the game' or starts out, he studies older hustlers to learn their techniques and may ask them to 'teach me some game'.

> Every dope fiend out there got some kind of game. Each one out there got they own style a game and by you hangin' tough with them you pick it up. They got some old timers, they got so old that they can't do too much. Then they see a young dude that they know play pretty fast, then they gonna give him some — maybe some pointers to make it more smoother.

Games are not rigidly drawn up. The best ones are improvised to suit the circumstances and revised as circumstances change. One important element is knowing the emotional makeup of the person whom one is manipulating as games usually involve 'playing on a person'. Since this is the case one must be constantly on the alert against being played on and the person doing so is looked upon negatively especially if he is a 'cut buddy' or crime partner and thus was placed in a position of trust. The game is known to be serious and one expression heard commonly when a person is busted or popped is "If you got to play, you got to pay". One minor aspect of the metaphor is reference to specific games especially card games in such terms as 'hold card', 'he played his hand', or 'to play a full hand against a royal flush' (to bluff with a loosing hand).

The general appropriateness of this symbolic usage is not merely a reference to the mental attitude of the player. It also reflects the highly competitive nature of these activities and hustlers' interaction with each other as well as the importance of strategy and chance in surviving on the street. It lends a particular character to relationships with others in which trust is difficult to establish and one avoids becoming emotionally vulnerable. On a deep level status is acquired by skill at playing the game. Justice and rights and obligations tend to take the form of reciprocal agreements which are weighted in the direction of the good hustler so that more people owe him than he owes people. Energy is spent convincing people that they do in fact owe him something.

One ex-pimp described this aspect of the hustle:

> It's the IOU — blackmail but in a different way. Just like getting one of your people out of jail, that's still part of the game and to get her out of jail to get her back on the street makin' money they would have to pay more because they got theyself put in that position to go to

jail but they couldn't get theyself out of it so they got to stick they hand out the bars and call they pimp and you got to come get 'em and the game you goin' to run to them is how much that shit cost for 'em to get out for the lawyers end.
It's part of the game she owe you because she got herself fucked up she couldn't get herself out so she owe you for that day on and she really don't know how much it is 'cause you never tell her. It's always you pay this, have this, on such and such a day and she have it and next week the same thing and a lot of times she might come up and say, I don't feel IOU this much. Then you just keep on.

The game metaphor is complimented by a less direct, but equally important theatrical or cinematic symbol complex. This generally refers more to the appearance of a person and the environment in which games are played. To 'fall on the set' is to make a public appearance. In most such situations one's status and style is subject to perusal and to 'have your act together' is to look good. 'Scene' refers to a particular type of activity along with its cultural customs and peculiarities. To 'dig the action' or 'check out the action' is to understand or find out what is happening in a particular set. At the center of this complex is a sense of not only play acting or pretending but a deep sense that one is participating in a movie. One consciously attempts to emulate this style and the movies have had a profound effect on the style of the street, especially such Black classics as 'Super Fly' which form a feedback system both reflecting and influencing the street. But on a more profound level many street men actually feel that they are participating in a movie. One informant described a particularly dangerous play as 'just like on TV'. Another, upon loosing his cut buddy, said, "I thought I was Super Fly but I wasn't shit. This is real, man."

At the root of these symbolic forms is the notion of image. The set or scene, usually a club or 'spot' or 'the stroll' or main street where 'everybody's hangin' out at', is a platform on which to display ones collection of illusory and real accomplishments. Partying and going to clubs, conspicuously spending money, are important activities in status maintenance. To 'keep up one's image' is a primary method for status acquisition and an important tool in hustling as well since money gets money. One's image is the face one wears to the world. It consists of appearance, rapping style, actions, and rep or reputation for past actions. It usually has real and false components.

Images are influenced by movie figures such as Super Fly and Humphrey Bogart as well as other Black entertainers and criminals.

This dude well, you think Humphrey Bogart had a fuckin' image. This mother-fucker, he make Humphrey Bogart look like mince meat or something ... We used to go out in these rental cars and go to this spot where my cousin hang out and these ladies hang out and where a few more sheisters hang out at and he'd fall in like he, oh man, like he Frank Nitty or some fuckin' body, you know and this the kind of image he had portrayed. He was just firm. He spoke what he said no matter what nobody said, he always was clean.

Being 'clean' or 'laid out', meaning well dressed in a fashionable way, is particularly important, as well as the way one expresses oneself. There is a variety of speaking types. 'Fast talking', usually in the context of a conning situation, is done to throw the person off. 'Smooth talking' is done often to women. The way one speaks influences the type of image one has. The following is a description of one image which the speaker regarded as effective but offensive:

There be some new faces on the set and instead of asking a lady, would she mind gettin' up and sit by him so he could tell her something, he'll tell her to get up and come sit here.

All of these manners of dress and action are described as 'style'. Style is, in a sense, the texture of an image. While it is possible to have style without money, money helps considerably. "Style is an individual thing. You would feel pretty bad if somebody was wearing the same exact thing ..." People with the best style are those who have developed a look which is unique to them. The style must in some way fit the person. Much energy is spent in developing a style. (Some men design their own suits and have them tailor made.) People have considerable pride when someone copies their own style, recalling particular times when this happened. On the other hand, there are certain parameters of taste which cannot be crossed particularly when a person is unkempt. High priority is placed on looking 'clean' even when one is 'strung out', if possible. One often hears "I wasn't no greasy dope fiend, I kept myself clean."

Another important aspect of one's image is one's 'rep' or reputation either in terms of physical prowess or hustling skill. A man with a good rep most likely will have an easier time maintaining his image. Often one can rely on reputation to see one through in a new situation.

I was big, man, people knew me all over town. Wherever I go somebody know me.

Rep also draws money:

Once people knew about me they used to send around for me to front they stuff (deal drugs).

A large part of the image is often illusory. One cynical ex-addict said that "to have an image is to portray something you're not but you're trying to be it. It's not honest." One creates these illusions not only by borrowing cars and clothes but by conning:

The way he used to get a chick to dig him was he used to fall up on the set and he might see me in the club and like call me and he say

> something like this, he say, "hey man, you heard so and so took off this thing today (robbed a place)?" "Yah I heard about it." I play into it, you know. Then he'll come with another thing, say, "Look here man, why don't you let me hold $300 and I'll get it back to you tomorrow or maybe before tomorrow." You know, he come up with this kind of shit to make this lady friend think well this is a hip dude and he got something going for hisself and his friend pretty hip too, so if she got a friend, she come over too and we get down.

When one informant was asked what makes a good image he replied, "Not gettin' bust at you game" or not having one's con exposed. Game feeds the image and image feeds the game.

Both game and image place an emphasis on exteriority, on one's appearance to others, the interior or hidden reality being irrelevant. To be slick or cool is to have minimized reactions which are strictly personal or emotional. One's real nature is very much a private affair. This is especially true of men. "A woman will give up her game to you. A man won't do that. If you listen to a woman, she'll tell you how to get over on her."

Control is another key factor, control of one's inner situation and outer situation as much as possible. Coolness is the ability to totally control oneself as this is something which is within one's power. Slickness is the ability to 'get over' on others, to slide over the surface of reality, to remain untouched by it. The barrier which intercedes between the inner and the outer reality is the image. Finally in the monotonous boredom of unemployment and poverty it makes life exciting. Games fill up time and have an air of glamour:

> It becomes like a game. It becomes like a game that real people are involved in, not chess pieces, or not checkers, not Chinese marbles but real people. A lot of people involves themselves in that trip more or less to keep life interesting. It's the limits that people go to ... A lot of people like to play with people. Probably half the population into playing with each other for the same reasons.

As with every people, so with street men, there is a discrepancy between ideology and practice. Status and slickness are effemeral and unstable. They are subject to pressures of economic hardship, law enforcement agencies, and ever-escalating demands for performance. Like the street money which supports the ethic, coolness and slickness are 'fast', they are acquired and spent very rapidly.

COGNITIVE AND SUBJECTIVE PLACE OF DRUGS

The meaning of drugs like the meaning of all other aspects of the street must be evaluated with reference to the game and the image. Finestone (1957) presented an analysis of the 'cat' in the fifties which described the 'kick' or high as an integral part of young unemployed male's street activities. The article seems in some senses as relevant now as it was then. He saw glamour, escape from boredom, and contempt for a system which doomed those men to low status as the foundation for the development of an alternate value system which included a high regard for drugs.

Among the men I interviewed these themes still predominated, especially a recurring emphasis on rebellion against an oppressive system, of resentment against the unique injustice of the ghetto. To most of these men the system has created a situation which forces individuals to break the law in order to survive. In addition, most have had some direct experience with people being 'hooked up' or sent to prison for a crime which they did not commit while others go free. There is an independent and in some senses transcendent sense of justice under which one man will 'take a fall' or go to jail for another and direct and swift justice is meeted out among peers. Using drugs represents a challenge and contempt for the system. It is "doing what I feel like" and ignoring legalities which stand outside the realities of the street, and have little to do with the business of survival.

Drugs provide challenge in themselves. They are exotic and glamourous, and therefore, exciting. They represent a new experience to be met and overcome. They are ritual aspects of initiation into the adult world of the game. They represent knowledge. They provide a knowledge or awareness of a unique kind since they are personal and subjective, outside the realm of social interaction or direct shared experience.

Drugs serve specific functions for the status of consciousness of the users themselves. The alteration of awareness (and secondarily, physiological state) dominates the meaning of drug experience. Social customs of use and pressures to experiment add depth and richness, but the "high is where it's at". This became clear to me when I was eliciting drug classification systems from a sample of heavy drug users. With slight variations all users organized drugs according to types of highs. Categories appear to vary with the user's experience, and some drugs have more variability in classification than others, particularly marijuana. In most cases the classification of types of highs was identical to the classification of types of drugs. Along with types, certain users order drugs according to strength.

The categorization which appeared most commonly for Black male users is illustrated in Figure 1. Some users worked from specific drugs to larger groupings from level 3 to level 2 then to level 1. Others worked in the opposite direction. In all cases the groupings were checked for completeness using common ethnoscientific methods. Some users classified drugs only to level 1 of complexity, others to level 2. Some users felt that all downs or ups could be substituted for each other while some felt that drugs could only be substituted within the level or grouping. Some classified drugs such as alcohol, marijuana and cough syrups in other categories. All referred to ups and downs in some way. Generally the high associated with downs 'mellows you out', 'makes you unaware', 'slows you down'; ups on the other hand, 'make you speedy', 'I can do things, you have a lot of energy'. Descriptions of psychedelic effects range from simple perceptual distortions to full blown hallucinations, 'you be out there, spaced out'.

For the sake of argument, I have isolated three drugs to illustrate my point with regard to the importance of subjective and cognitive aspects of drugs. These are the most potent of the drugs in each of the level 1 categories, heroin, cocaine, and LSD. The cognitive status and meaning of these drugs is by and large differentiated by the effects of the high on the execution of street activities.

Heroin is either classified as a down or 'in a class by itself'. It is considered to be the 'heaviest drug on the scene'. One indication of its status is that people often refer to it as simply dope or drugs. Other common names for heroin are stuff (alternately used as a sexual term), skag, doogie, monster, H, etc. (The list is virtually endless as documented by Lewy & Preble, 1973). Perhaps one of the most telling expressions for heroin is 'king hero'n' for in the street heroin is the king of drugs. It is the most challenging and, in the long run, the most expensive. The language and customs surrounding heroin use are more highly developed than those associated with any other drug. An entire image, 'the dope-fiend image' is ascribed to its users. One reason for this is of course the fact that it is highly addictive and requires much time for maintenance, but another is the unique high. It is a down, and as such has a mellowing effect.

Figure 1. Classification of street drugs

Level of complexity	Drugs					
Level 1	Ups		Downs			Hallucinogens[1]
			(Opiates)[2]			
Level 2	Cocaine	Speed or Amphetamines	Heroin	(Drugstore drugs)[2]	Downs or Barbiturates	LSD
Level 3[3] (specific drugs)		Methamphetamines Dextro-amphetamine sulfate or (Giz) Desoxyn		Blue morphine Dilaudid Cough syrup	Yellow jackets Reds Trues Quaaludes	

1. Often users with little psychedelic experience attempted to fit this into other categories. Acid is often reclassified as an up.
2. Opiates and pharmaceutical drugs are sometimes absent.
3. The distribution of marijuana is variable. Usually it is listed as a 'down' or hallucinogen. Alcohol, when it is considered a drug, is believed to be a down.

One user described it as a feeling like "you know when you're dreaming and you start fallin' in a dream and you wake yourself up, but slower". Another described it as a sensation of heat rising in the body. In addition, it has a peculiar effect on the mind. It seems to stop the stream of consciousness for certain users. For most it dulls the emotions.

> With heroin you don't have no feelings. It's a relaxed state. The world floats right on by. Nothing seems to affect you.

It relieves anxieties about problems.

I: You can tolerate a lot of stuff that ordinarily you wouldn't tolerate. You got any problems, just make you forget anything that ever existed.
A: If you had a problem, would you remember it?
I: I would remember but it wouldn't be any problem. To me it wouldn't be no problem, not as long as I was high. I wouldn't be concerned about it.

These effects of heroin are especially significant when one considers the problems faced by street men. Generally there is a great deal of interpersonal tension in their life situations, problems with money and jobs, the law, their women. These problems become a constant preoccupation and their potential solution is usually complex, extending over a long time. There are usually extended periods during which nothing can be done about a particular situation.

> Time is dead when money is tight, when people are occupied elsewhere — working or in school. Time is dead when one is in jail. One is doing dead time when nothing is happening and he's got nothing going for himself (Horton, 1972).

A wide variety of tactics are used to defend against this peculiar mixture of anxiety and boredom. Alcohol and marijuana are used as a distraction, but heroin offers a special kind of relief:

A: How does heroin affect your sense of time?
I: Heroin makes time non-existent.

Problems that seemed pressing lose their significance.

> It takes away pain. Most people take it to cover up what got you depressed.

With regard to the street code of coolness, this is significant. Heroin is the ultimate drug of detachment. It is difficult for a young man, who is not a developed hustler and who is generally facing a large number of 'hassles' to remain cool. Ironically, coolness is what is required to handle the situation. Heroin provides access to coolness, both in that it is the king of drugs and thus the ultimate challenge and in that the high makes the person feel good, detached and able to cope with his problems. But the attitude toward heroin is ambivalent due to its addictive power. This is expressed in such terms as 'tragic magic', 'white lady', 'white whoe', and 'monster'. Many successful street men never use heroin or only briefly experiment. Indulging in heroin, "letting your problems get the best of you and wanting to cover up" is an indication of a weakness in one's image. The ideal of using heroin is to try it and to let it go, to in fact conquer it by not having to use it.

Cocaine is radically different than heroin. Sometimes called ice, it gives a numbing cold and hyperactive sensation and is popular as a sexual drug. "It makes you feel like a stream running through your body, a very cool stream". Not only does it literally give a cool feeling, but is socially and psychologically cool:

> Coke, I sorta can control myself. It'll mostly keep me up. It have me up I'm aware, really, that paranoia, you know. You can sorta feel somebody watching. You always looking around, always on the go . . .
> I always categorize cocaine in the street code as being a money's man drug 'cause I found that cocaine let me come, left my head cool. Seemed like I could just think more clearer. It make me have more guts than I usually have, that's why I call it a money's man drug. Every hustler I know probably uses it, probably snort it. I used to feel like a king, I had extra pep.

Cocaine is thus an aid to the hustler in that it promotes awareness of the environment to the point of paranoia, and encourages an aggressive and active attitude. Coke also "deadens the inside", minimizing feelings so that "a person use cocaine project a very cold image. Most people use cocaine project a very cold attitude." Like heroin it facilitates detachment, but apparently in a slightly different sense. Most people emphasize that heroin deadens anxieties and removes a person from a sense that he is in the world, thus indirectly minimizing concern for others. Cocaine does not cut down anxiety but directly eliminates concern for others and sympathetic emotions. It thus is an excellent drug for manipulators and people who play on others.

Cocaine is cool and fast literally and symbolically. It is a status drug not only because of the qualitative properties of the high but because "its very expensive and it don't stay no time". The high which results from shooting lasts 15 to 20 minutes. It is not unheard of for a person to shoot $100.00 worth of coke in a single evening. It is a drug of glamour, extravagance, and conspicuous consumption, especially suited to fast money and a successful hustling image. In addition, it increases the ability of men to prolong intercourse thus supposedly increasing the users appeal to women.

These two drugs complement each other providing interior and exterior detachment. They are often used together in a 'speed' or 'speed ball' which allows one to be both down and up simultaneously. Heroin "takes the edge off coke" and coke prevents heroin from "putting you out of it". Qualities of each drug are manipulated to "put you just where you want to be", attuning the consciousness to the demands of the situation.

The importance of subjective effects of drugs in facilitating maintenance of the street code is clarified by the negative example of LSD or acid. LSD is little used in the ghetto. It is considered to be a white drug and all the men I interviewed had copped acid from whites. In addition, the high is considered to be radically different from the commonly accepted street drugs. While other drugs allow contact with ordinary reality to some extent, LSD trips may entail periods where ordinary reality is completely absent. Few men I interviewed had any experience with it and those who did regarded their use as experimental and limited. Many had tried it once and refused to try it again, stating, "I don't fuck with that acid, it puts you out there, makes you crazy", or "you're escaping from reality into another dimension". "It makes you where you see things you know, you see things that aren't there. You imagine shit. Acid you think you're somethin', a bird or something gonna fly."

References to flying are frequent. It is common for people to report and emphasize lack of control on acid. Any experienced acid user will tell you that to have a good trip you must "give in to the experience, flow with it". Most street men regard this as exceedingly dangerous. Stories about acid commonly refer to suicides and flashbacks:

> I know a friend and he was high off of acid and he was on the top of the roof. Think he was Tinkerbell and he could fly and he jumped and killed hisself. What I heard that you take acid today and a year later you have flashbacks and go into another trip, and you could've just maybe taken an aspirin or something.

The reputed unpredictability of acid and the danger of flashbacks at undesirable moments keep many ghetto men from trying acid or duplicating their first experience. The acid high is grossly impractical on the street:

> When I started getting addicted I felt like when I got off (shot up) I could go out whereas couldn't with acid, no. You can't go into a place and say, "this is a stickup" and whatnot. You start laughing or you might try to shoot yourself in the head instead, who knows. You don't have no control and stuff with hallucinogenic drugs and I think that most people on the street who's into the drug scene, I don't think they fuck around with acid before going out, and doing a roll (committing a robbery).

An acid trip may last anywhere from 8 to 36 hours. This too is impractical since during that time one may trip out or hallucinate in "another dimension" and thus be unable to go about one's business. Subjectively the acid experience is antithetical to control and external orientation and thus unsuitable for life in the street.

Drugs, then, are evaluated with reference to their utility or fit with the street game. To be up or down is to allow one's consciousness to slide on a scale of intensity which suits the situation, adds interest, and facilitates in handling situations. Drugs add another dimension to life experience as well as complimenting the game and the image. In addition, the stuff game is very lucrative.

DOPE FIEND RITUAL

Repetitious behavior surrounding drug use can only be regarded as ritual in a very general and secular sense. William Burroughs (1953) repeatedly pointed out that there is nothing sacred about heroin or indeed about street use of drugs in general. Overriding cosmologies and sophisticated philosophical systems are rare.

For most, reality is a simple matter of survival, in keeping with traditional American conceptions of a single reality which, while it may be viewed in different ways by different people, is essentially the same and is always there to be seen. To "tell it like it is" is to see this reality clearly, an operation which is regarded as entirely possible and appropriate. Beliefs in the occult, while present, are overlays to this conception, often being extensions of the material realm. For most street men I spoke with, visions of mystics and acid users are in the realm of the crazy. Hallucinogens just don't fit the way the world is.

The symbolic content of shared customs is not regarded as terribly significant. What is significant is the concrete results or effects of any action, not its reality status. Most street men play with the world but their object is realism or the creation of something that looks real rather than symbolic expression. There are a few exceptions to this. For example, the subtle and sophisticated use of language in activities such as sounding or ranking (ritual insults) expresses and elaborates competitive interaction between the men. Vocabulary describing drugs and activities is also symbolic as noted a number of times above.

But the actual texture of street activity is "dealing with reality as it is". This is perhaps what is so fascinating about street customs. They are predominantly pragmatic and rational with little room for symbolic elaboration. While one often hears symbols and rituals in primitive society explained in terms of "that is the way it is" or "our fathers before us did it thus", a dope fiend can almost always give a practical explanation for anything he does.

Standard strategies for dealing and copping are designed to avoid ripoffs. The presence of a number of armed guards at a large deal is obviously related to protecting the sale. Leaving one's money in the car with a companion is another. Fronting drugs avoids exposure. Placing stashes at various locations on the streets avoids possiblity of being charged with possession and sale. Shooting half a bag first limits the possibility of overdose. Booting (repeatedly injecting and re-injecting) cocaine increases the high. Shooting up in the bathroom provides access to water and a natural shield of privacy. Other locations avoid detection. Even the symbolic language facilitiates a certain degree of secrecy, as evidenced by certain men's unwillingness to discuss ways of talking about drugs which are used in the presence of straight people, often involving code words or expressions which are only detectable to those who have knowledge of the particular scene or those preoccupied with drug use. Such usage is of course symbolic but is very different from the kind of multivocal and condensed sacred symbols typical of most rituals.

While there are many problems in describing street behavior in ritual terms, as is clear from this discussion, there are a few special areas, I think, where one can productively talk about young urban Blacks experience of drugs in the rarified vocabulary of the symbolic anthropologist.

First, it seems to me that the type of shared experience induced by traditional ritual processes is adequately provided by the sharing of the drug high. Victor Turner (1969) has described this particular kind of consciousness and social interaction as existential or spontaneous communitas,

> Essentially, communitas is a relationship between concrete, historical idiosyncratic individuals. These individuals are not segmentalized into roles and statuses but confront one another rather in the manner of Martin Buber's "I and thou" ... (131-132).

One component of communitas is "liminality, structural inferiority, lowermost status, and structural outsiderhood" (134). Comradship and equality are other aspects, "it is rather a matter of giving recognition to an essential and generic human bond". Street drug use does not follow exactly all of the formal definitions of sacred communitas but there are some important similarities. First, there is the low status of drug users in the eyes of the rest of society. Second, there is the element of acceptance. "I always liked to be around dope fiends. They accept you. There's no put down. They can't speak about it, because they junkies theyself."

Drug users share a common vision of the world. Users like to get high in each other's company. There is a great deal of social pressure to do so, especially when the individual who offers resistance has been known to use in the past. It is not uncommon for a refuser to be offered a 'hit' a number of times. One ex-addict reported:

> When I go back there, and don't get off, they think hey man something wrong with you, you crazy or something?

Especially in 'dope houses' it is common for users to 'sit around and bull-shit' or rap while enjoying the high.

Another element of sacred ritual is often the sharing of indescribable experiences. It is perhaps for this reason that symbols are employed.

They are a medium for indefinate and personally experienced awarenesses which are impossible to codify. Drug experience, as was mentioned above, has a similar quality in that the experience is impossible to transmit during the course of ordinary social interaction. Drug users share knowledge impossible to share with non-users, thus having a special bond similar to the bond created in the ritual context.

This shared knowledge has a special importance for the adolescent providing a substitute for and symbolically representing ritual initiation into adulthood. For the young man, learning about drugs is an important part of learning about the street. The fact that a high cannot be explained in the course of ordinary social interaction adds a special feature of initiation into a kind of knowledge. As in ritual initiation, there is a shared experience of revelation (of a new type of awareness or quality of consciousness passed from initiate to novice). Psychologically, for the young man on the street this experience is very important. Most of them have few models of successful strategies for entering adulthood. The formulas of their elders are often outmoded and illsuited to the stresses of the urban scene. What was a good approach before integration of life in the city is not applicable. There is a peculiar hunger for knowledge among these men which is not satisfied by an unrealistic education system geared for youngsters with completely alien cultural backgrounds. This hunger is often satisfied by those who are easy to reach, peers and men from immediately older age grades especially.

Perhaps, in a sense it is also the form of the situation which is sought after and valued. Like ritual initiation, it is the symbolic revelation of what is hidden. On some deep level it parallels the revelation of obscure adult experience and strategies for "making it in the system". More directly, drug use is an entree into other knowledge about the street game.

One young man's experience aptly illustrates these points:

> I was square as this room man, I didn't know nothin' and wasn't up on females too much either then until I started hangin' tough in the streets. Using drugs, that right there started me really gettin' in touch with a lot of different other sources, gettin' hip to the game, knowing different angles and holes out there in the street. Being related into that drug field, hey you get up on a lot of things — money-wise and whoes, all different type a things that you won't run into ordinarily. Experiencing some of those things out there you get the feelin' that you're really havin' fun.

Drugs are obscure to those who are not involved with them. Users often report that they never knew how many people were using a particular drug until they started using it themselves. Then, miraculously, it seems, a whole group of people with similar experience come into focus:

> People I grew up around, that was what they was into. The neighborhood that I stayed in was a drug traffic neighborhood. Everywhere you went they have different types of drugs. They was there all the while when I was comin' up but by me not bein' aware of what was goin' on I never knew until they was aware I was into it.

This is facilitated by the fact that drug use is, by legal necessity, a relatively secret activity. Users will not often use in front of squares for fear of being 'popped' or arrested and there is a sanction against use in the presence of children. For example it is not uncommon for drug users to have never seen a person shoot up until trying it themselves:

> I was scared even to take a shot, man. I've heard of it but I never really seen somebody do it. People used to walk around and say, hey Jack, people around stickin' needles in they arms and I really didn't believe them because I didn't know whether they had shit out there like that. I wasn't hip to nothin'. I was square as hell until I saw this person do it and I saw the position he was in and how he was feelin' and I said, you know, fuck it, I'll try it and that's when he hit me.

Other aspects of the drug scene make it appealing. The fact that these young men, "out to experience the world", regard new adventures as a challenge to conquer in the same way that gang fighting competition and activities are challenges. In the case of heroin this challenge outweighs the danger and takes the form of the rationalization that "I won't get strung out, it can't happen to me". Since the drug experience is difficult to explain, young people's questions concerning what the high is like are usually answered simply by "it feels good, man", arousing curiosity and interest. In addition, a man new to the street may be experiencing stress in establishing his image and looking for means to bolster it or may be finding it difficult to "be cool". These factors combined with the elements of the thrill of secrecy and immediate social pressure to join in use form a cluster of converging vectors which make initial use highly likely.

The body of concrete knowledge concerning drugs, dosage, types, prices, etc. provide palpable information which the new initiate goes about acquiring. Often, this knowledge ends in the realization that drugs also have negative consequences, as one's experience includes observations of overdoses, adverse reactions and addiction. With heroin, the knowledge of addiction, that one is sick because of drug use is often transmitted from old addict to new addict:

> I rationalized it. I always kept myself up and looking down on junkies as raggedy and I just couldn't see myself there. Three, four months later, the first jones I ever had, I thought it was the flu and took off from work that day. I was sittin' down in my car and this guy come

up on me and I let him sit down and I was talkin' with him and he was a user and he told me, I didn't have the flu. And I say,"well I know I got the flu, the way it affect me" (because see, a jones affect you just like the flu). He say,"got $10.00, I bet you I can tell you how that flu go right away". I thought it was game. I say,"hey man, if you need something you don't have to game on me. I give you $10.00 and you can get off." He say,"no man, well give me $10.00" and I gave it to him. He brought the drugs back so I snorted it and right away it went away, that's when I first became knowledgeable about being sick. Seem like that made it worse because it brought out the real dependency. Not knowing this normally, I probably would have stayed home a few days, probably it would have gone away.

This revelation often ends the period of initiation for the heroin user and his career begins in ernest.

SUMMARY

An entire complex of supporting dynamics on the street contribute to the young Black man's vulnerability to drug use and ultimately to the unsuccessful integration of the experience into a viable lifestyle. Drugs are part of a street ethnic which demands an extraordinary level of detachment from an unusually stressful environment. They provide states of consciousness which meet this requirement, and having fun is fulfilled by substances which cognitively represent status, glamour, slickness and challenge. Internal and external pressures to continue after initial experimentation are great.

At the same time drugs become an added expense in an already difficult financial situation.

Beyond this the simple cognitive focus of law enforcement agencies on young Black offenders makes it more difficult to engage in drug related activities in anonymity.

In the final analysis, the essential nature of the subjective drug experience complements the lifestyle of the street.

NOTES

1. The research for this article (1975) was conducted under the auspices of the National Institute of Drug Abuse at the Center for Theoretical Social Research on Drug Abuse, Division of Addictive Sciences, Department of Psychiatry, School of Medicine, University of Miami, Public Health Service Research Grant No. DA01073-02.
2. Note on language: Throughout this work I have quoted interviewees' actual street language without censoring vocabulary. I feel that these words are so

much a part of street life and language that to alter or eliminate them is to misrepresent the texture of that reality. Obscene words are not obscenity on the street, but expression.

BIBLIOGRAPHY

Burroughs, William, 1953, Junkie. New York: Ace Books.

Finestone, Harold, 1957, Cats, kicks and color. Social Problems, 5(July):3-13

Hannerz, Ulf, 1969, Soulside: inquiries into ghetto culture and community. New York: Columbia University Press.

Harner, Michael J., 1973, Hallucinogens and shamanism. London: Oxford University Press.

Horton, John, 1972, Time and cool people. In Thomas Kochman (ed.), Rappin' and stylin' out: communication in urban Black America. Urbana: University of Illinois Press.

Keiser, R. Lincoln, 1969, The vice lords: warriors of the streets. New York: Holt, Rinehart and Winston.

Lewy, Marc G., & Edward Preble, 1973, Tragic magic: word usage among New York City heroin addicts. Psychiatry Quarterly, 47(2):228-245.

Liebow, Elliot, 1967, Tally's corner. New York: Little, Brown & Co.

Stephens, Richard C. & Duane C. McBride, 1976, Becoming a street addict. Human Organization, 35(1):87-93.

Turner, Victor, 1969, The ritual process: structure and antistructure. Chicago: Aldine.

MERCEDES CROS SANDOVAL

9

PATTERNS OF DRUG ABUSE AMONG THE SPANISH-SPEAKING GAY BAR CROWD

1. INTRODUCTION

The drug problem in multi-ethnic and fast-growing Dade County had reached alarming proportions at the end of the decade of the sixties. At that time, several preventive and treatment programs had been created aimed directly at combating the abuse of drugs. It was obvious from the beginning, that the Spanish-speaking population (considering its concentration and density in the overall county population) was failing to participate in any substantial numbers in any of those programs. It was assumed, therefore, that this lack of participation did not mean that there was no drug problem in the Spanish-speaking community, but rather that the established drug programs had failed to appeal to the Spanish drug abuser because they were not responsive to the characteristic cultural configurations of the Latin community. Therefore, a grant establishing the Spanish Drug Rehabilitation and Research Center was approved and the program became operational in late 1972. The objectives of this center were the study of the Latin drug experience, and the delineation of the different drug subcultures which coexisted within the Latin community, in order to arrive at an understanding of the Latin drug addict, his lifestyle and his needs. It was hoped that this study would lead to the design of a treatment modality which would take into consideration the unique needs and characteristics of the Latin drug abuser. Such a treatment modality would be in good consonance with the Latin cultural configuration and values; as well as with the resources, assets and liabilities of the Latin Community. In this manner, it would help expedite the drug abuser's adaptation to his cultural and social milieu here. The success of any treatment, after all, can only be measured in terms of degree of effectiveness with which the rehabilitated will function within his community.

As Director of the research component of this program, I visited the drug programs already operating in Dade County, as well as the jails, and other such institutions which served or somehow dealt with drug abusers. I interviewed those drug abusers who volunteered information hoping to better understand the nature of the drug problem among the Spanish-speaking population in Dade County.

2. METHAQUALONE ABUSE AMONG THE SPANISH-SPEAKING POLY-DRUG USERS WHO ATTEND THE GAY BARS

In June of 1973, I first discovered evidence of the heavy abuse of Methaqualone among the young poly-drug users who participated in the Gay Bar subculture. The first contact took place at the Women's Detention Center (County Jail). I visited this center on a weekly basis after having asked the social worker in charge to call me whenever there was a resident charged with a drug-related problem. After a while it became apparent that the majority of women incarcerated for drug offenses were federal prisoners, the so-called 'mulas' (mules). These were women who came from South America to this country for a short period of time and for the purpose of discharging their drug cargo. Most of them were not drug users themselves and did not know anything about the drug problem in Dade County. Some of them claimed that they did not even know exactly what their cargo contained. They stated they were induced into the drug traffic because of economic pressures in their home country and the relatively high reward offered to them by the dealers at relatively low risk. Most of them were hired for one or two trips (so as not to arouse the suspicion of the authorities). They hoped to gather two or three thousand dollars in these operations, which would enable them to raise their economic position back home.

Discouraged by these contacts which yielded little information on the drug problems of the Spanish-speaking population of Dade County, I asked the social worker in charge of this Women's Detention Center, to set up interviews with me with any Spanish-speaking women in detention regardless of charges. In this way, I met an 18 year old girl who had been convicted for prostitution and sentenced to 20 days. We met together several times while she was in jail and after she was freed. Based on the rapport established, she became one of my key informants and one of the persons who gave the green light to my acceptance by the drug users who attend the Gay Bars.

The second member of this Gay subculture I met was also incarcerated. This Cuban male was serving a short sentence at the Stockade (County Jail), on a robbery charge. After a few interviews, I acquired additional information about the Gay subculture and its drug problem, this time from the pusher's point of view. It is interesting to note that neither of these two informants is actively homosexual. Both hang about with the Gay crowd because they find support and fun among them. The girl, however, had a long-standing love affair going with a homosexual boy.

After intensive interviewing, I asked these two informants if they would care to write up a small essay in which they would describe their lifestyles, daily routine, life expectations, social support systems, etc. These essays (see Appendix) were very useful since they answered several questions which the interviews had raised. They also gave rise to other relevant questions such as the need to further investigate the various functions that drugs fulfil in the lives of these young abusers; the relationship, if any, between homosexuality and the use of methaqualone; the values, morals, and supportive systems within this population; and the family background of group members. I hoped to draw a profile of these young drug abusers which would give us a better understanding of their problems and would be helpful in developing a treatment modality better suited to serve their specific needs. I also wanted to develop more effective ways of reaching them. Another goal was to discover common denominators, besides their drug problem and their homosexual preference, which might be important variables to consider when studying and understanding the pathology of the Latin community. Examples of these variables could be some specific deficiencies in their homelife, educational exposure, or economic security. If any such variables were found to be significant, they would have to be highly considered by any treatment or educational program designed to combat illicit drug use among the Latins.

3. METHOD OF APPROACH

During the months of August, September, October, and November of 1973, I became a daily visitor to the homes, boutiques and drop-in-places of the members of the Gay subculture. In total, I interviewed thirty-two drug using members. I also rapped and talked for long per iods of time with other members of this subculture who did not use drugs, but who gave me valuable information as to their lifestyles, values, etc. By 'Gay Bar sub-culture' I am referring to Spanish-speaking gay persons who frequent certain bars located in Dade and Broward County. They also frequent several beach spots in Miami Beach and boutiques and other shops which specialize in the kind of attire sought by the group.

The first interviews were set by the informant I had met at the Women's Detention Center. These contacts in turn, led to many others. Most of the interviews took place in the late afternoon and evening since many of the members of this subculture, especially those who abuse drugs the most, get up late. The interviews took place primarily in the homes or places the informants called home at the time they were contacted. At least fifteen of the informants, the heaviest users, were living in a sort of communal setting by renting two or three apartments clustered together after abandoning their parent's home. This residential pattern enabled them to share resources (cars, food, drugs) with greater ease. Two such unstructured communes were frequently visited. Other interviews took place in the homes of lovers and friends where the drug users were staying for the time being. Still others took place at the beach, in the boutiques and the cafeterias they frequented in their daily or weekly routine.

All of the interviews were face to face ranging from one to three hours. Also, short raps with two or more potential and former informants were fostered whenever possible as a means of engaging in group discussion. These allowed an opportunity to double-check specific points of information and also encouraged the shy to open up for more intensive interviewing. Contradictions and misleading statements were clarified under challenge by peers. Thus, more information regarding their lifestyle was obtained. At least nine of my informants were interviewed approximately six additional times. As a result of this, my preliminary open-ended interview schedule was modified to gather information which would better reflect their problems, value system, and behavioral patterns. Thus, questions were included regarding their sexual preference, their attitudes and opinions about the Gay crowd, about the way they viewed the Cuban family as compared with the American family, about their religious preoccupations etc.

I used the tape recorder whenever the informant showed any positive interest in being taped. Some loved it and let their lives and experiences flow while being taped. The information was transcribed immediately and the tape erased to avoid any breach of confidentiality.

4. FINDINGS

4.1 The Gay Bar scene

The Spanish-speaking Gay subculture in Miami cuts across social, ethnic, racial, and economic lines. It is centered around the Gay Bars, as well as the boutiques and beach spots most favored by this crowd.

Among the most popular bars at the time of this study were (Aug.-Nov. 1973): The S, The AT, The L, The M, in Miami, and The K in Hallandale. The C, The M II and A were also frequented by this group.

AT was patronized by a gay crowd with a ratio of three males for every female. On weekdays, an average of thirty-five people showed up: during weekends, the crowd soared to an average of two hundred and fifty to three hundred. S was patronized by both straight and gay people. On weekdays, a small crowd of forty attended this bar. On weekends, three hundred to three hundred and fifty people gathered there. The L primarily catered to tacky female gays. On weekdays, a crowd of thirty was seen there; on weekends it reached a hundred. On Tuesdays and weekends, Miamians swelled the clientele to over two hundred who gathered at The K in Hallandale.

In general, the atmosphere of the bars is somewhat bizarre, a world of cheap fantasy, loud music and striking colors. Most admired are the extravagantly dressed. The dealers of downs, while resented by some, generally are the center of much attention and are sought after by the crowd. These down dealers are by no means rich; they just manage to make a living. The owners and managers of the bars do not deal in drugs; they apparently only realize profits from alcohol sold at the bars and from the admission fees. Pills and other drugs brought by the patrons manage to get to the bars. Some of the bar clients are not drug users or use them very occasionally. They attend the bars sometimes reluctantly and then only because these are the only places where the gay crowd can socialize in an atmosphere of acceptance. These clients, making up as much as 70 per cent of the crowd, occasionally will take some drugs, but only as a means of getting a 'nota' (high), to socialize with more ease. From the informants I gathered that possibly only thirty per cent of the customers are active seekers of drugs or dependent on them.

The habitual clients of the Gay Bars call themselves modern, meaning that they are in the midst of what is 'in', the vanguard of new trends. Within this classification, however, some discreet categories can be recognized. The 'Exquisitos' (Exquisite) who exhibit fantastic wardrobe and seem appropriately dressed for a costume party, attract most attention. Their expensive and extremely shocking apparel boasts such brand names as Christian Dior, Yves St Laurent, and Oscar de la Renta, which are the envy of many of the clients. Goodie shoes $ 60 to $ 80 a pair, are also extremely admired. This attire is the product of relentless hustling,[1] shop-lifting, and continuous barter and exchange. In this group it is not a disgrace to borrow from somebody else, sometimes indefinitely and without permission of the owner, items of clothing or adornment. Most of the members of the 'Exquisitos' group are the Drag Queens[2] and their competitors and admirers, the Fag Hags.[3] Among these 'Exquisitos' a very snobbish soft slang is used within the word 'bello' (beautiful), 'divina' (divine), 'me fascina' (it's fascinating to me), 'fantastico' (fantastic), 'es una locura' (it's madness), and of course, 'exquisito' (exquisite) repeated every other word. The 'Exquisitos', exhibiting very narcistic behavior most of the time, are so outlandish that they are shockingly funny and outright 'goofy'. Some of them have some artistic talent, like the Drag Queens who star in the Gay Bar show and who are amusing and entertaining. However, some of them look like somewhat decrepit clowns, especially the day after a party when, in broad daylight, they parade on the beach suffering from a hangover, only half-painted and dyed with remnants of their evening make-up. In general, they enjoy getting the attention of everybody. This is especially apparent when, dressed as females, they take off their tops to show off their bare and unyielding breasts. They obviously find great enjoyment in the surprised, or even better, shocked response from spectators. Sometimes they are so young that one cannot help thinking that they look like children wearing their mother's wardrobe. They like to mimic popular singers and entertainers with their brash talent. Their greatest joy comes from showing their numerous photograph albums, mute witness to their prior fashionable exploits. They love to dress in costumes and to surprise and confuse you with an unrecognizable disguise (sometimes dressed as females and other times as males). Often they brag about the expensive clothes they wore and the great time and 'tono' (high) they had on the occasion recorded by the photograph.

A second identifiable category within the Gay Bar subculture is that of the 'Exoticos' (exotic). Also wrapped-up in their 'trapos fabulosos' (fabulous rags), the Exoticos seem to have more sober taste than the 'Exquisitos'. Their claim to fame is their elegant 'jet set' taste as compared to the blatantly outlandish attire of the 'Exquisitos'.

In general, the persons who wear or hope to wear 'trapos fabulosos', with their red make-up around the eyes, their neat, very important, hairdos and flashy expensive 'zapatacones' (high-heeled shoes) are pennyless. Many of them came from formerly well-to-do Cuban families. They seem to have found congeniality and company with the children of people with whom their parents would never have associated in Cuba. The basis for their alliance rests on the common ground of their sexual preferences and their liking for extravagant clothes. Some of my informants, no doubt, would have channeled their exhibitionist needs and their liking for expensive clothes into the protective but shallow atmosphere of middle and upper class social clubs, had they lived in the Cuba of the fifties.

The 'Supercool' are the sexy set who wear small tops and very tight, sensually-designed and provocative clothes. According to some informants, most of the dealers dress in this fashion.

The 'Riff-raff' are more of the beat set, poorly dressed according to their income but mostly expressive of their own taste for a more simple if somehow rugged outfit.

In general, the philosophy of life of the habitual client of the Gay Bars is very present-oriented and extremely hedonistic, emphasizing immediate experience and pleasure. A strain of shallowness, irrelevancy and superficiality permeates it. Rivalry, suspicion, competition and gossip take a heavy toll, adding extra pressures to the intrinsically tense atmosphere of the Gay subculture. Direct experience is most valued; indirect experience is worthless. Emotional and sensual experiences are more valued than academic or intellectual ones. Sex and money to spend on clothes, especially on shoes, are the immediate goals of their lives, which by no stretch of the imagination, is planned or programmed for a period longer than two weeks. Their lives are a series of loud and intensive sensual experiences, ephemeral emotional relationships, marked by the slow devastating passage of time . . .

The habitual clients of the Gay Bars tend to be completely out of touch with the demands of the technological and achievement-oriented society. They are, for the time being, drop-outs.[4] One way or the other, the Gay subculture helps its members cope with the problems of adjustment arising primarily from their sexual preference or lack of it. The extravagant atmosphere of the Gay Bars provides some of its members with the only stage possible, supported by an admiring audience to which they might relate. Here, the effeminate homosexual males and the not-so-gay but vain female Fag-Hag, enliven the fabulous setting in which the socially unacceptable gay like to roam.

4. 2 The special study group

4. 2. 1 Sample characteristics — In the special study group of thirty-two habitual bar clients, twenty of the informants were male and twelve were female. Regarding sexual preference, seven of the males reported they were Drag Queens; five were non-transvestite homosexual males; six males considered themselves bisexual; and two males called themselves heterosexual or straight. Seven of the females reported themselves as homosexual; four called themselves bisexual; and one thought of herself as straight. However, three of the females who qualified themselves as bisexual reported they preferred sex with Drag Queens. All of the informants were single, except for three males who had been married and had already obtained a divorce.

The informants' age ranged from fifteen to twenty-eight years old. The great majority of them were between eighteen and twenty-two years of age. Entrance to the Bars seemed relatively easy, since many minors made their way in with little or no problem.

All of the informants were of Cuban extraction except for a boy from Puerto Rico and another one from Salvador. Among the former, all were born in Cuba except two who were born of Cuban parents in the United States.

Twenty of the twenty-eight informants who were born in Cuba came to this country accompanied by their parents. Among the eight who did not, two were reunited with their parents in less than six months; four lived with their grandparents for a period of less than two years, and one was reunited with his father after spending five years in an orphanage. One who came with friends in a boat eight years before, had lived alone since then. Sixteen of the informants came to the United States when they were seven years old. The rest came between the ages of eight to fourteen. Thus, most of them had been in this country for at least thirteen years and their formal education had taken place here. All of them reported having problems arising from the value-conflict between their Latin cultural background and that of their host country. Even though all were bilingual they considered Spanish as their primary language.

One third of the interviewed population still lived, at least theoretically, with parents; another third lived alone with friends; while the remaining third shifted from their parents' to friends' and lovers' homes and vice-versa.

Ten of the thirty-two informants came from homes broken by divorce, separation or death of one of the parents.

The parents' age was not a significant variable since it varied amply. The comparison between the former and present occupation and status of the breadwinner at home was not a significant variable, either, since it also ranged across the steps of the occupational ladder and only demonstrated the erosion and changes in status which have been characteristic of the acculturation process undergone by the Cubans in Dade County. All but one of the informants were very favorably impressed with the qualities of closeness and warmth they thought was characteristic of the Cuban family. Five of them, however, reported that the Cuban family did not allow for individual independence and claimed it was "full of gossip". Twenty-eight of the informants, even though they were not planning to have children, reported that if they had any, they would rear them strictly. Fifteen of the informants claimed that the members of the American family did not care much for the welfare of the other members and that the American children altogether miss the closeness and warmth of the overprotective Cuban family.

The majority (twenty-eight of the informants) felt that Cubans were friendlier, happier and brighter than the Americans. They also reported that Cubans really 'know how to have a good time'. They perceive Cubans as great dancers, conversationalists, as having good social manners, being more cosmopolitan, and not as socially awkward as the Americans. Reputedly, Cubans also knew how to dress in good taste. All but one of the informants reported that 90 per cent of their friends were Cubans.

Seven of the informants had regular jobs in offices and stores; five were fulltime students still supported by their parents; three made a salary as stars in the Gay Bar shows; four were male prostitutes; three were female hookers; three were pimps and hustlers; two exclusively dealt in drugs; and five reported doing nothing but hustling (including selling drugs when available).

All of the informants had a high school diploma, except three who dropped out in ninth grade. Three had finished the first two years of college.

All but one of the informants believed in a very personal God. None, however, expressed any fear of this God ever punishing them for what in the larger society qualified as antisocial behavior. As a matter of fact, many expressed great concern about God and felt that they were on good terms with Him since God did not consider either their drug abuse or homosexual practices as sins.

Opinions varied in regard to illegal activities, other than drug use, in which many of them engaged. One third of the population objected outspokenly to these activities. The rest either were indifferent or actually found the engagement in illegal activities as a status achieving exploit. Nine of the informants had been arrested for drug possession, three of whom had also been arrested for breaking into homes, and two for assault. Four informants had been arrested for shoplifting, and two had been arrested several times on charges of public drunkenness. The most common crime committed by the informants was hustling, shoplifting, passing false checks and drug-dealing.

The informants, according to their drug behavior, can be categorized following H. Stephen Glenn's classification (1972) into the following groups:

— Two 'experimental' users who tried out various drugs to find out about them without following a consistent pattern.

— Five 'social recreational' users who took drugs in social and recreational situations. They did not actively seek drugs; they used them only when available.

— Three 'seeking' users who sought out drugs and connections to obtain them. These latter viewed drugs as an essential part of their social and recreational activities. Some depended on them more often than in a social context.

— Fourteen 'dysfunctional' users. Drugs already dominated their lives, and they were dependent upon them to be able to cope. Drug abuse was so much a part of their lives that their livestyle had substantially changed and their ability to maintain a job, keep up with regular school schedules, etc, was already disrupted.

— Eight 'suicidal' users. Their life style and their health had been so affected by the use of drugs that behavioral patterns of self-destruction seemed apparent. One of these characterized as suicidal recently died from smoke inhalation he suffered when caught stoned in a fire.

4.2.2 Range of drug use and functions of the drugs — In the timeless world of the Gay Bar's scene, drugs serve multiple adjustive functions.

Drugs help the members of the Gay Bar crowd cope with the problems arising from their sexual preference, erasing sexual inhibitions which could have restrained the homosexually or bisexually inclined.

Drugs also serve to facilitate social interaction. Shy people become overtly aggressive under the excuse of 'no te pongas brava conmigo porque estoy arrebatada' (don't get mad at me because I am in a frenzy), which allows for a displacement of guilt and a comfortable disassociation from one's disapproved behavior by blaming it on factors beyond one's control. In general, informants said that drugs allow for better, smoother social interaction, especially cocaine and methaqualone. With all inhibitions cut off, one feels 'fenomeno' (phenomenal); one thinks that one is a 'superfly'; powerful and independent. Drugs, thus, can give the user a great sense of mastery. It becomes obvious, when observed, that for some the use of drugs is a means of acquiring status. Those known to take a lot of pills without loosing control are 'hot tunas', 'estan en algo' (they are 'in') — who feel or reputedly have a great sense of mastery.

Drugs also give the users a feeling of identity and a common bond, in contrast to the lack of cohesion of the straight non-drug-using crowd. Thus, the drug-using crowd, which shares common values and prejudices, acts as a strong supportive system for its members.

Among those hard-core users, drugs can be the tools of self-destruction. This seemed specifically true among some of the transsexual males who had gone through the sex-change treatment. Two of those interviewed were dissatisfied with the change. They felt no identity with either of the two sexes and heavily used drugs with what I believe to be an unconscious desire to overdose.

Most informants attributed different effects to the different drugs but agreed that all of them affected individuals differently according to their personality, their emotional state, their health and other such circumstances. Nevertheless, in general they felt that 'tecata' or heroin was to be feared and heroin users avoided and ostracised. They thought it was highly addictive, provoked social withdrawal and was dangerous since there was no way one could be sure of quality or strength of the heroin being sold. The majority expressed actual terror when confronted with the thought of using a needle to get a high. They were aware and very afraid of the infections and other side effects that heroin usage might bring about.

At the time of this study heroin went for $10.00 a bag. My feelings are that this group of poly-drug users in general was neither sophisticated nor aggressive enough to have dared to use heroin. Most of them lacked the pharmaceutical knowledge necessary to survive in the heroin subculture. Also, most lacked the risky hustling skills necessary to support a heroin habit.

Twenty-two of this population of poly-drug users reported having used heroin at least once. Eight had only snorted it a couple of times, while four had shot it with some regularity. Those who had used it regularly reported that 'tecata' gives a 'great high', an unbelievable 'rush'. At the time of the interviews only two of the informants were taking heroin with some regularity, but neither was addicted to it. However, eight of the informants reported having been addicted to heroin in the past, for periods stretching from five months to two years. Of the eight, five were females who were prostitutes at the time of their dependency and obtained supplies of heroin from their pimps who dealt in it. Double-checking showed at least three of them to have been very dependent on it. Three males claimed having been hooked on heroin. One of them, while on heroin, engaged in all types of criminal activities to obtain the money for his $110.00 a day habit. I met this informant in prison, and he had a long record of breaking into homes, car stealing, assault, and drug dealing. In one of his numerous periods of incarceration he went 'cold turkey'[5] with no medical assistance. At the time of the interview he claimed to be dealing only in methaqualone, other downs, and grass. Another male reported having kicked heroin addiction by substituting it with grass and downs (this same treatment was used by three of the female informants). In this regard the medical literature reports (De Alarcon, 1969) that methaqualone has been prescribed for insomnia in drug patients, particularly heroin addicts. It is described as providing a 'jolt' or a positive pleasant response in those patients for whom it is prescribed.

In general, however, this Gay Bar crowd seemed very afraid of heroin. Many of them reported that the Puerto Ricans are the ones who use it in the Spanish-speaking population of Dade County. One of the males who had been addicted explained that he was hooked in Puerto Rico, 'pais de escapistas', (Puerto Rico, country of escapists). He explained that heroin which causes isolation, anxiety, and antisocial behavior, has a negative reputation in this Miami group where extroversion, acting out, and goofiness are so admired and rewarded.

Psychedelic drugs had been amply tried by this Gay Bar crowd. Among them organic mescaline seemed to be the most popular. Peyote and psilocybin were also around. Mescaline sold for $2.00 a hit as did also dimethyltritamine (DMT). Under the popular names of Sunshine, White Paper, Window Pane and Angel Dust, some of the psychedelic tablets were sold at $2.00 each. Acid was not very popular and about 60 per cent of those interviewed claimed they did not like it. None of them chose it as their drug of preference. The few who liked it called it a mind expander. One of the informants claimed that under its effect, while tripping, he was capable of striking the moon against the sun, which gave him a great feeling of mastery.

Specially among those informants who used downs and still atended school and a steady job, amphetamines, 'uppers' or 'speed', were more popular than acid. Their lives were an 'up and down, and down and up' sequence of experiences; a down at dusk and an up at dawn cycle. Under the popular names of Black Bird, Black Beauties and White Crown, a number of amphetamines were sold at $1.50 a tablet. The brand names that circulated most were Eskatrol, Bramadez and Ritalin hydrochloride.

Cocaine enjoyed a great reputation. Only a few of the informants had not tried it, and of those who had tried it, only a few did not care for it. Cocaine had status. It was highly regarded and considered an expensive, 'classy' drug used by the sophisticated and rich set. The reason the members of this crowd did not use it more often was its high cost. They considered it harmless and non-addictive. Among some of the members of this group it had the added reputation of being an aphrodisiac. Besides, cocaine excites you and that is why it is called 'perico' (parrot), since the person under its influence cannot stop talking and acting out. This behavior is considered most appropriate especially at parties. The user who becomes the life of the party with that uncontrollable power to act out serves as a catalyst of social interaction and is most accepted and admired. Most of my informants took cocaine when they attended parties and social orgies footed by middle-aged Cuban businessmen and by younger American executives who in this manner participated in one of the less dignified aspects of the Cuban culture. At such parties, cocaine was offered free in small trays. Girls were invited to actively participate in the sexual theater called 'cuadros'.[6] Some of the males attended as passive spectators when invited or as hangers-on with their invited female friends. Sometimes they also actively participated in these 'cuadros'.

According to my informants, regular cocaine sold for $ 25.00 a spoon. Better quality cocaine sold for $ 40.00 a spoon, and pure cocaine went for $55.00. A spoon yields enough for several 'pases'.[7]

Marijuana was used by most of the informants, oblivious of the bad name and low status it used to have among Cubans of only one generation ago. In general they considered it harmless and thought it was great to cut inhibitions. Marijuana indeed enjoyed a well-established reputation as a safe drug. Neither high nor low status was associated with it. Many complained that due to its peculiar odor it could not be used in the Bars, since it could be promptly discovered by the management and the police; but they used it often in private parties, at the beach, and in their homes. At the time of the interviews, Jamaican grass sold for $25.00 an ounce, Mexican grass varied from $30.00 to $35.00 an ounce, while Colombian grass, reportedly the best available, sold at $35.00 to $40.00 an ounce. Another hemp derivative drug which was popular was Tetra-hydro-cannabinol (THC). Many of the informants also used Amil-nitrate (poppers) with some regularity even though it lacked prestige and reputedly was only used by children and those users who were desperate and would take anything.

Methaqualone was the common drug of abuse. Even though the informants have been characterized as poly-drug users, methaqualone was their drug of preference at the time of the interview. This was so because of its availability, its price, its safe quality, etc.

Methaqualone is reportedly relaxing and aphrodisiac. It has the virtue of reducing inhibitions, and allowing the 'closed closets'[8] to open to a life of greater sexual satisfaction. According to one of the female abusers who was most dependent upon them, "they make me feel happy, free, 'goofy', relaxed and 'loca y a mi me gusta estar loca' (crazy and I like to be crazy)". In this manner she expressed her satisfaction at being able to free herself of all social controls and inhibitions. The drug literature refers to the alleged quality of methaqualone as a reducer of inhibitions and as an aphrodisiac in the following way: "While the drug lowers inhibitions, and increases sexual desires, it actually lowers the ability to perform sexually" (Gamage, 1973).

Most of the methaqualone abusers had no fear of it, since they firmly believed it created no dependency. They made this claim right in the middle of a 'tantrum' provoked by the bad news that none were available that evening.

Among the methaqualone drugs, the most commonly sold were Quaalude 7-14 (Rorer), Sopers or Sopor (Arnar Stone) and Parest (Parke & Davis) which were selling for $2.00 a pill. Canadian Quaalude (reportedly underground laboratories)[9] sold for only $1.50. Today (Spring 1975) all of these pills sell for $3.00 each.

Other downs or barbiturates were also on the black market; such as, Tuinals, Seconals, and Nembutals which sold for $1.50 or $2.00 a pill. Today (Spring 1975) they sell for $3.00 a tablet.

Most of the informants obtained supplies from the dealers interviewed. However, a lot of them obtained them from Cuban doctors and drug store employees who were their friends and who would give them prescriptions or provide them with refills at no extra cost. The pharmacies sold one hundred Quaaludes for $30.00, which would enable a user to keep half for himself and deal the rest with a potential profit of $100.00. This allowed them, if the right contacts were kept, to support the habit and still make some profit. In reality, this was not such a businesslike enterprise. A lot of the drugs frequently changed hands without the exchange of money. In other words, customarily, when one user had a large supply, he would share with his friends as an investment for the times when he had none available. Even the dealers who made their livelihood selling drugs were not rigid. They often gave their friends free drugs.

A lot of drugs made their way into this subculture through prescriptions by the so-called 'pill doctors' and through those written by doctors who performed the sex-change treatment. The great majority of 'pill doctors' were non-Cubans who issued prescriptions after the $10.00 appointment fee was paid. More often than not these 'pill doctors' failed to examine their so-called patients. On the other hand, the sex-change physicians, also non-Cubans, had legitimate reasons to medicate with these pills. Examples are cases in which the patients were undergoing silicone and/or electroysis treatment.

According to my informants there were around fifteen to twenty pill dealers in the bars, each serving around fifteen clients. These dealers obtained their drug supplies mostly from Puerto Rico and also from the free access they had somehow obtained to doctor's prescription blanks. Others had contacts with drug store employees who were known for their 'laissez-faire' practices. Dealing in downs never seemed a very profitable business transaction. It is my impression that few physicians engaged in this traffic except for some of those retired 'pill doctors'. As for pharmacists, it seemed that some were casually careless and could have pocketed extra dollars from their refilling practices. I did not obtain conclusive information regarding the seemingly inexhaustible sources of downs which made their way into Dade County from Puerto Rico. Neither did I obtain any lead as to the sources of the so-called 'Canadian Quaaludes' which reportedly came from underground laboratories. Tighter security on prescription pads would certainly reduce the traffic. A closer watch on the refill practices of the pharmacies would also help.

Nine of the informants interviewed reported having overdosed at least once while using downs. Five of them were treated by friends who gave them enemas. The treatment also included forced, relentless walking so they could not go to sleep. It also included strong coffee and attempts to induce vomiting. Unfortunately, a few months before this field work took place, one of the members of this gay community overdosed with methaqualone and the treatment that time proved ineffective.

Quaaludes were used separately, but, as in the case of grass, they also mixed well with almost any other drugs. Eighteen informants favored taking grass and downs together. Six had often taken downs and heroin together. Ten reported taking downs and speed regularly in a sequence, as a means of coping with the demands of their jobs and school. Many of the informants mixed Quaaludes with alcohol or with Cuban coffee to make it 'blow'.[10]

Even though the thirty-two informants at the time of the interview were taking methaqualone habitually, not all of the informants reported methaqualone as their drug of preference. For instance, seven informants preferred cocaine, but they did not use it regularly because of its high price. Three preferred marijuana which they took daily, mixed with the downs. Two preferred heroin but were very much afraid of getting hooked. Among those who preferred downs, most of them liked Tuinal better; one preferred Nembutal; nine definitely expressed their preference for methaqualones. The remaining nine informants expressed an undifferentiated preference for any type of downs. All of the informants were habitually taking Quaaludes because of its availability, its price, and the fact that it does not require a great knowledge of pharmacopeia since it is standardized. Furthermore, it has a reputation of being a non-addictive, safe, clean drug with high status.

4.2.3 Sequential pattern of progression of drug use — Five sequential patterns of the progression of drug use seem to be apparent:

1. Eleven of the informants started with marijuana when they were around fourteen years of age, proceeded experimenting with cocaine, acid, speed, downs, heroin and returned to downs again. Two of those included in this group never touched heroin.

2. Eight informants started early with marijuana, then experimented with acid, speed, cocaine, some heroin and finally downs.

3. Five of the informants started with marijuana, then went to Quaaludes, acid, speed, cocaine, heroin and returned to downs.

4. Four informants started with marijuana, proceeded to use cocaine, downs, acid, speed and then returned to marijuana and downs.

5. Four informants used marijuana, then cocaine, acid, speed and then downs.

CONCLUSION

This study has described the heavy abuse of methaqualone by the Spanish-speaking poly-drug users who attend the Gay Bars in Miami. It has suggested the way drugs function in their lives. It has also attempted to portrait the lifestyle and value system of these young poly-drug users. A basic assumption has been that better understanding of their needs and problems allows identification of possible sources of support which could be tapped for rehabilitation purposes.

This gay crowd's inordinate need for recognition, admiration and acceptance has been noted. Also acknowledged has been the manner in which methaqualone reduces inhibitions and facilitates the release of acceptable behavioral patterns which express and satisfy their needs. It has been suggested that the social rejection gays experience is one of the most important causes of their maladjustment. Consequently, it is only through membership of this subculture that their identity needs and sense of belonging are fulfilled. However, it was observed that this subculture's world-view and value system are limited, confining, and tinged by the paranoid glasses of the rejected. It is dysfunctional in that it hampers the coping capacity of its members and their ability to adapt to the wider social context.

In view of these facts, it seems that any rehabilitation program aimed at helping these drug abusers would have to focus on a special type of counseling that is sensitive to the problems of the gay. One of the objectives will be building another accepting support system which would in turn make them less dependent on their peers. It is possible that families could become an alternative system of support if the shame and ultimate rejection of its gay members were either alleviated or terminated.[11] For that purpose it would be necessary to work on developing a family-therapy model designed to bridge the antagonism between the members of this population and their parents. This strategy seems advisable if one considers that many of these informants expressed great admiration and loyalty toward the Cuban family, even if not accepted by their own. The dichotomy created by their expressed positive feelings toward the Cuban family lifestyle and its rejection of them has resulted in generating a great amount of guilt and shame. I have credited much of the acceptance and trust shown me by many individuals in this group to my age. I feel that as part of the parental generation, they regarded me as a substitute parental figure, whose non-evaluative or non-censuring support satisfied their need for parental acceptance.

If a family therapy model were designed for this population, this should be considered. Counselors of their same age might provide them peer approval which although needed, does not meet the requirements of the much sought-after parental approval. Counselors of their parents' age will be more trusted and respected and could serve as mediators between the gay and their parents. This need for approval from authority figures was inherent in the answers given by some of the informants regarding their strict child-rearing beliefs. Apparently, this suggests that part of their maladjustment may have resulted from the conflict brought about by the permissiveness of the American culture, where most of them had spent many years being socialized, and from the authoritarian cultural background of the family.

It is pertinent to point out that during the brief period I worked with some of my informants, I was able to establish enough rapport to encourage six of them to engage in therapy. The counselor to whom I referred them also belonged to their parents' generation. Therapy in one of these cases led to a rejection of drugs and better adaptation to a mainstream lifestyle. Another informant's sense of self-esteem was improved to the point that he reduced his drug intake and began to hold a steady job. The other four discontinued treatment shortly afterwards. My feeling is that had I persisted in my personal supportive efforts on their behalf the latter might have continued in therapy. A point for consideration, therefore, is that intensive outreach might serve to successfully engage many of them in treatment. In a sense they are easily located because the places where they hang around are well known. My feeling is that such outreach programs should be implemented. Those special counselors which could be considered by gay drug abusers as substitute authority figures could be very effective in helping them resolve their many interpersonal and/or intrapsychic adjustment problems.

SUGGESTIONS FOR FURTHER RESEARCH

1. A study of the patterns of drug abuse among the American Gay Bar population and a comparative analysis of those findings with the ones presented in this paper.
2. A study of the patterns of drug abuse by Cuban 'Straight Bar' clients and a comparison of those findings with these.
3. A comparative study of the use of methaqualone in different subcultures and subgroups.

APPENDIX

Daily routine of a drug abuser

"I am eighteen years old and I came to this country when I was only three. When I was seven my mother, a habitual drinker, abandoned my father and left with another man. For years, me and my brothers resented my mother for having abandoned us. I have no feelings for her anymore.

I was reared by my grandmother who moved with us when my mother deserted us. I started using drugs when I was fourteen. Then, I was attending P Jr. High school. I was given a cigarette of marijuana which I smoked by myself by a lake. Then I started trying other drugs. By the time I was in the seventh grade I regularly smoked pot and also took acid and downs. I liked to take drugs to be 'in the in-group', which consisted of both Cuban and American kids. While attending tenth grade in High School, I started using heroin. I used it only for five or six months and then I quit because 'I saw what it was doing to my friends'. I also tried cocaine which I got from older Cubans whom I dated. I have been taking Quaaludes regularly for the past year, I take six or eight every day.

I get up around 2:30 p.m. I go with my friends, either shopping (shoplifting) or to the beach located in 21st Street and Collins Avenue. Since I don't have a car I always call somebody to pick me up. At the beach there is a lot of gay people. Even though I am not gay, I like to hang around with gay people because they are warm and understanding: 'they take you where you are at, they ask no questions'. When I get to the beach, the first thing I do is find some Quaaludes. There is always somebody selling them because after they go to the doctor and get their drugs, they all go to the beach. Most of the time they give them to me for free, other times I have to buy them at $2.00 a piece. When I get them, I take two. We stay at the beach until around 5: p.m. Then we go home, C. and I. At home we take a nap, around 10:00 p.m. we go to the bars, then I take three Quaaludes sometimes more. Almost all my friends take drugs, mostly Quaaludes. Some of them take acid when it is available, others who can afford it take cocaine. Some of them take Junk (heroine), but only once in a while. None of my friends are hooked. Sometimes we take 'poppers'. It is something that comes in a bottle and you smell it. It gets you a high similar to the one you get with glue. People are always passing it around. Different friends of mine take different drugs. My friend S. always has downs and so does R. and M. and sometimes D. If they have a steady job, they don't take drugs regularly, but on weekends when they go to the bars where one can have a lot of fun. After the bar closes we go to private parties and get more drugs. Since I am not working now, I go to the bars on weekdays to get a high. I usually go home at 5:00 a.m. and then go to sleep.

Most of my friends are gay. I hang around mostly with Cubans. Straight people also go to the bars, I think that they are very smart since they don't let anybody take advantage of them. Gay people are all 'fucked up'. Some of them are trying to change their sex. Even those who have changed their sex are still 'fucked up'. La., for instance, took an overdose and died because after having changed sexes she still was not happy. La C., for instance, is always so depressed that she has no desire to live. They really don't know what they want out of life. Some of them have a steady relationship with a man and they are happy for a while but most of their relationships are not too lasting. They spend all their money buying clothes and jewelry. When they don't have money, they have to do a lot of shop-lifting. Once a friend of ours was caught shop-lifting at B.'s department store and the police made a deal with him. They let him free in exchange for an album with pictures of the gay people and since then, the gay people have been more persecuted at B.'s departm store. Also, the gay people have to spend a lot of their money for the chan change of sex treatment, including hormone shots, electrolysis and silicone."

NOTES

1. 'Hustling' — e.g. "I had to hustle very hard for my bread" — refers to numerous activities, both legal and illegal, such as shop-lifting, stealing, drug peddling, begging, bartering and primarily prostitution in which many of the members of this study group engage as a means of making a living without having a steady job. Hustling also refers to fast living.
2. Drag Queens are the transvestite active males.
3. Fag Hags are the extreme feminine females who seek social and sexual relations with Drag Queens and with gay males.
4. Sometimes I wonder whether or not the use of drugs was the catalyst of this situation. When considering them individually, I could not but toy with the idea that back in Cuba they would have dropped-out anyway without the use of drugs, turning into the numerous dependent members of the over-protective extended family. Some of them, however, I felt were pressured into the use of drugs by the forces operating in this society which speed up the process of adulthood and mastery of full responsibility over one's life. For the latter, I felt that, if given the opportunity and the time they needed to mature and accept responsibility, they could eventually join the mainstream of society.
5. 'Cold turkey' is the process of withdrawing from drugs without any medical or psychological help; as opposed to detoxification where a great deal of medical and/or psychological support is offered to the individual.
6. 'Cuadros' (frames, pictures) are the sexual performances, circus or theater in which three or more participants engage in various sex acts.
7. 'Pases' (pass, turn) is the term used for the nose inhalation or sniffing of cocaine.
8. 'Closed closets' is the term used for the heterosexual with hidden homosexual tendencies.
9. The literature: Gamage (1973:4) states that methaqualone is not manufactured in underground laboratories, I have no knowledge of the sources of these so called 'Canadian Quaaludes'.
10. The term 'blow' is used for the activated effect of a drug when mixed with alcohol, coffee or any other substance.
11. A similar recommendation to the one offered by the author in March 1975, has been defended in an article in Tropic Magazine, September 28, 1975, called 'Parents of Gay'.

BIBLIOGRAPHY

De Alarcon, R., 1969, Methaqualone. British Medical Journal, 5636, (January): 122-123.

Gamage, James R. & E. Lief Zerkin, 1973, Methaqualone. Report Series, National Clearinghouse for Drug Abuse Information. Series 18, No. 1.

Glenn, H. Stephen, 1972, Some basics of drug abuse. Unpublished manuscript. Regional Training Center, University of Miami.

FRIEND OR STRANGER

The members of a sub-culture, such as the ones discussed in the previous section, require identification of membership. The fact that they are using drugs, that this use is illegal, and that there is always a danger of infiltration by police, necessitate that the members of such a set be able to identify outsiders.

In certain situations the fact of knowing a person or a whole group may act as safeguard, while in other contexts common ethnic background, being gay, or having the marks of a user may distinguish friends from strangers. The latter of course may be undercover agents and it is important to avoid such persons.

Carlson and Page report on two different situations in which strangers are identified and even prevented from understanding what is being communicated. In the first chapter heroin users in Honolulu must prevent being busted by following a set of behavioral rules by which addicts can be distinguished from strangers. The second chapter considers the use of a street language, called 'pachuo', by the cannabis users in San Jose, Costa Rica. As was the case in the former chapter, the attempt here is to keep outsiders of doubtful confidence from understanding what was being communicated. Once again we are dealing with ritual actions and ritual language.

KATHERINE A. CARLSON

10

IDENTIFYING THE STRANGER: AN ANALYSIS OF BEHAVIORAL RULES FOR SALES OF HEROIN

Heroin addicts in America have been described by many researchers, among them Agar (1973) and Stephens & Levine (1971), as comprising a subculture. This subculture is predicated on addicts having a shared system of specialized knowledge. It is to be expected that behavioral rules used in the addict subculture will reflect the necessities imposed by a particular environment and by a particular lifestyle. The insecurity of much of the addict's existence, the day to day unknowns of an adequate drug supply and continued personal freedom, must be considered in any analysis of the addict's actions. These general insecurities of maintaining his habit are countered by the addict's belief in his own control over occurrences which are liable to threaten this maintenance. He has confidence, for example, that he can control the outcome of various encounters if he adheres to certain rules and procedures which are part of his subcultural knowledge.

This belief in control is well illustrated by the reliance of many addicts on 'conning' as a means of obtaining money and drugs. There are also other situations where control is important. This paper deals with the situation of an encounter between the seller of heroin and the prospective buyer when the buyer is a stranger. The procedures which are followed under these conditions are directed towards the reduction of the uncertainty of the encounter's outcome, in this case, whether or not the stranger is an undercover narcotics officer and will subsequently 'bust' (arrest) the 'dealer' (the seller of heroin). These procedures have a very regular application and inspire considerable confidence in their practitioners. Further, their organization varies in a systematic way according to certain circumstances. Such procedures are thus forming a system of behavioral rules which contain elements of pragmatic logic, ritual behavior, faith, and intuition. This system is referred to in this paper as Rule Set 1.

The information on which this paper is based was collected during a year's fieldwork with heroin addicts associated with various treatment programs in Honolulu, Hawaii. Most addicts were associated with the Waikiki Drug Clinic or the John Howard Methadone Maintenance Program. Length of addiction varied from a few months to more than twenty years, informants' ethnic affiliation included Hawaiian, Japanese, Chinese, Filipino, and both local and mainland Caucasians. Information was collected through informal interviews, casual conversations, and observation of interactions within the treatment environment and in neutral settings such as coffee shops.

In the course of this data collection, twenty-two addicts directly discussed some aspect of the encounter between a dealer and a stranger. All but one of these individuals were or had been themselves dealers of heroin, an occupation which was also practiced by the majority of all informants. These sales may be of large quantities on a regular basis, of small quantities regularly, or of small quantities irregularly. Dealing, at whatever level, is thus part of the repertoire of almost every addict, and thus almost every addict is concerned with sales to a stranger.

Addicts try to maintain their addiction despite considerable opposition not only to their drug use itself but also to the ways in which they support it. The opposition which is a major problem for addicts is that of the police department, for they may be busted for both the possession and sales of heroin, and also for the variety of other illegal ways in which they make their money.

The main efforts of law enforcement are directed towards stopping the sale of heroin and arresting the dealer. One method used by police to obtain such arrests is the use of undercover agents who pretend to be addicts or trustworthy non-addicts. The undercover agent's purpose is to obtain proof of selling activity by making actual purchases from dealers, or by observing such purchases. The addict is aware of this emphasis and knows that if he is dealing, he is the primary target of the undercover agent. Although dealers have other problems and threats besides that of being busted for selling, it is the potential of being busted which most concerns them, and which they make the greatest effort to avoid. There are many risks associated with dealing, and not the least of these is the likelihood of an arrest on a felony charge. The addict persists in selling despite these risks because dealing is the most efficient and reliable way to support a habit.

The addict copes with the pressures of his environment in a rational way from his perspective of desired goals. In order to maximize his pleasure from heroin, and minimize the risks and negative aspects involved, the dealer will attempt to exert some degree of control over the likelihood of getting busted for sales. Part of what one must learn in order to be a dealer then, includes procedures through which one is able to protect oneself.

> Dealing is easy, anybody can do it. You just set yourself up with some stuff and the hypes will come to you. What's hard is not getting caught, because they're always after you. That's where the skill comes in and why most dealers don't last very long; they don't know who to trust. That's what you have to learn, you just get to know who you can trust, get a sixth sense for when something isn't quite right.

This statement by a dealer is a typical generalization of what addicts say they do to protect themselves. Most of my informants were able to extend this general statement to involve a discussion of the following procedures which form Rule Set 1.

The most important element in knowing who to trust is the dealer's ability to identify the stranger who seeks to buy from him as either another addict, a 'safe' non-addict, or an undercover agent. In the way in which it is being used here, an 'addict' is anyone who uses heroin, whether or not they are actually physically addicted. A trustworthy or 'safe' non-addict is someone who does not use heroin but has a legitimate reason for wanting to buy some. This includes people who are non-addict dealers, something of a rarity at the street level, and people who may buy for their friends or mates who are addicted.

The position of the stranger is a difficult one for the addict to resolve. A stranger, when a sale is involved, is anyone whom the addict does not know it is safe to sell to. The 'stranger' then, can be a person whom the addict knows, has seen around, or is otherwise familiar with, but whom he does not know **not** to be an undercover agent. Because Honolulu, and especially the district of Waikiki where many addicts are dealing, is an area of much transiency, there are always a lot of people who are strangers to a given dealer. Somehow the dealer has to find a way to respond to these people which will protect him from being busted by a stranger who is also an agent.

There is one basic precaution which is available to the dealer which will protect him from the possibility of selling to an undercover agent. This is given so frequently by dealers, and even by addicts who are not dealers, that it can be considered as a rule for proper behavior.

> I have nothing to do with strangers; I'd never sell to them. If anybody got busted because they sold to an undercover guy, then that's a stupid guy. There's no excuse for that. Even if a stranger comes in with a guy who's a friend, the dealer won't sell to the stranger. He'll sell to the other guy and say, "you sell to the stranger."

As outlined above, and in almost every other discussion of the stranger, this rule, Rule 1, is "Never sell to strangers". As long as this is followed and the addict sells only to those individuals whom he personally knows to be addicts or acceptable others, his chances of being busted for sales are minimal. Even if there is an 'informer' (an addict working for the police) who approaches the dealer for a sale, he is still protected by not selling in the presence of a stranger.

The informer plays an important part in many busts by introducing the agent to the dealer in an attempt to circumvent the prescription against sales to strangers. Technically, the police can also use the informer's testimony as sufficient evidence by itself to convict a dealer, or supplement this by having the informer buy with marked money which is then recovered from the dealer. However, I have never received any mention of these technical possibilities as a problem or a threat in discussions with addicts, and this method apparently is little, if at all, utilized in Hawaii. It may be that convictions are too hard to obtain by this method or that informers are reluctant to so publicly reveal themselves. The informer was frequently mentioned, however, as the individual who set up the bust, as illustrated in the following account:

> There was a guy I used to deal with in High School. I hadn't seen him for a while and the guy had turned informer. He set me up with a narc. I'd told my friend before, to get the money from this guy and bring it, and then he could sell to the stranger. But my friend brought the stranger to me anyway. I hadn't been dealing long, but I knew enough not to sell directly to a stranger. But like a fool I sold to him. I was nervous about it, but I trusted my friend.

By following the rule of never selling to a stranger, the addict has minimized the risk of dealing, but he has not maximized his return. For there are many strangers besides the undercover policeman who would like to buy from him: beginning addicts, out-of-town addicts, addicts looking for a new 'connection' (dealer), and so on, including previously mentioned non-addicts. The danger of sales to strangers is that the undercover agent is usually a stranger. However, that there are other strangers, who have a legitimate claim on the dealer's services, and who can provide him with what may be needed income or a means to a 'fix' (shot of heroin), makes it difficult to practice a caution which in principal seems ideal. So, although most addicts will state as a rule that one does not sell to strangers, and that they themselves do not sell to strangers, they are also able to talk about exceptions to this rule: times and conditions under which sales to strangers may still be permissible.

> You never trust people you don't know pretty well, because people who trust strangers get busted. I don't think you can really pick out another junkie; even if a guy looked okay, I still wouldn't trust him. The only time I'd sell to a stranger is if he'll do it in front of you. If the guy refuses to, then there's something wrong. Or if you've seen them using before, then you'd believe it if somebody says they're allright.

Even the addict who is quoted so emphatically on page 103 that he has "nothing to do with strangers", continues on to cite the way in which such a sale should be conducted.

> Even if he's sick, there's no way you'd sell to somebody you don't know. Now maybe it's different, but then, if a total stranger told me he was sick, I wouldn't believe him. I wouldn't know if he actually shot dope. But if I was going to help that person, I'd make sure he shot it right in front of me.

According to statements like these, it is possible to avoid the restriction against selling to strangers providing certain actions are taken which will continue to give the addict some degree of protection from a bust. Rule 1, "Never sell to strangers", represents a cultural ideal. Like most cultural ideals, actual behavior does not precisely conform. The prescription against selling to strangers is logical, but impractical. Since the major reason for not selling to strangers is the probability of one being an undercover agent, a test which by some method would eliminate that danger would provide a good alternative to the avoidance of all such sales. The addicts quoted above have mentioned an act which does this, and this can be referred to as Rule 1a. The prescription now being followed is "Never sell to strangers unless you have seen them use". This permits sales to unknown addicts, but effectively blocks sales to agents who, by definition, cannot use.

Sometimes, however, for various reasons, this actual observation of use is not required by the dealer. In such a situation, an additional rule is applied, Rule 1b: "Never sell to strangers unless someone you trust vouches for them". In most cases, vouching for a stranger means telling the dealer that you have seen them use, and thus the stranger is not an agent. Used this way, Rule 1b is simply an indirect version of Rule 1a. But a stranger can also be vouched for as a trustworthy non-addict.

If you could as an addict really trust other addicts, Rule 1b would be essentially equivalent to Rule 1a. In the bust described on page 104, however, reliance on someone else's word proved to be inaccurate. There are many pressures on an addict besides those of friendship and mutual trust, and these pressures may, and frequently do, result in the addict working as an informer in exchange for some favor from the police.

Addicts who themselves have been busted on a lesser charge than sales report that they were offered their freedom in exchange for introducing an agent to their dealer. Most addicts reject this offer, some accept it with the intention of not following through, and some become informers. Knowing this, the addict also knows he is taking a greater chance taking someone else's word for the stranger's identity than if he were to verify it himself. The use of Rule 1b thus represents an increase in the amount of danger which is involved in the sale over what is present in Rule 1a.

The addict has still another way in which to assure himself of the stranger's identity and calm his fears about selling to an undercover agent. This is actually a number of different methods which share certain similarities. These may be used when neither Rule 1a nor 1b has been applied, or in addition to either one of them as an extra precaution. I have grouped these methods together as Rule 1c: "Intuitive methods; identification of the stranger through clues of dress, talk, and behavior." These rely primarily on the addict's previously mentioned 'sixth sense,' his intuitive ability to distinguish an undercover agent from a true addict.

The proper addict is expected to be able to recognize narcotics' agents whenever he comes in contact with them. This is an integral part of addict folklore and was related to me by many informants. That some of these same individuals had previously been busted for their apparent failure to do this did not appear to contradict the basic recognizability of the police. These nine of my informants who had been busted for sales, and others who had come in contact with agents, all claimed to have been suspicious about the stranger's true identity.

It is obvious that what the addict refers to as his 'sixth sense' is in reality clues about a person's identity which are for the most part not articulated or consciously recognized. But when these clues are presented inaccurately, or are not present for someone who claims to be an addict, then the dealer begins to feel that there is something wrong with the identity as it is portrayed. When pressed, some addicts are able to articulate some of the clues which they use, but these are mainly only partial explanations of the entire identification process.

One frequently cited clue used under Rule 1c was that of dress. The agent trying to pass as an addict can be recognized because he dresses differently than a regular addict. One individual was discussing a charge pending against him for a sale which he claimed not to have made. He reinforced this denial by saying that "the guy who's supposed to have busted me was wearing desert boots and an aloha shirt hanging out to hide his gun." This stereotype was immediately recognized by other addicts present as being a description of an obvious agent, and since no one would sell to such a person, the charge was indeed false.

Another informant claimed that he initially refused to sell to an agent who subsequently busted him for possession "because he just didn't look right." A request for clarification of what not looking 'right' entailed resulted in this description: "He was a big guy, looked more like a football player, like a 'fat weasal'. He was wearing a t-shirt and grubby jeans . . . had short hair and a thin mustache." This description is quite different than the previous one, and in fact, the descriptions of inaccurate dress cover the entire range of possible attire, suggesting that it is not so much the clothing itself which provides the clue but the attitude with which it is worn. The above addict summed up his description with the conclusion that "he looked like he was trying to pass for something he wasn't."

Addicts also look for the type of 'hustle' (way of making money) used by a stranger, and whether or not it is illegal. Involvement in illegal activity indicates that the individual, if not an addict, is at least unlikely to be a policeman. One woman was busted by an agent who promoted himself as not an addict but as a small-time hood, and convinced her to sell to him for resale to others. An older male addict once gave me a series of ways in which one would decide whether or not to trust a stranger:

> If a guy will sit down and use with you, that's the first thing you look for with a stranger. If he won't, then be suspicious. And then look at the things he does, how he gets money. It's important that the money come illegally. It's hard to be a hype and get money legally . . . And then there's the way he talks about his experiences, where he's from, what he's been doing.

According to this man, the most important and first way of evaluating a stranger is to see him use, the application of Rule 1a. Additionally, he applies Rule 1c, by looking at hustle, and also by judging how the stranger talks.

Perhaps the most important clue utilized under Rule 1c concerns talking, both the choice of words and the sorts of things which are discussed. Addicts place a premium on the ability to manipulate others verbally, and this is a skill which is especially present in the dealer. This ability is also a necessary attribute of the undercover agent. One agent was accepted by an addict despite considerable suspicion because "He talked like an addict, like he just didn't give a damn. And he used certain words, certain terms for dope, like an addict would."

I have further received several examples of encounters where the agent was able to convince the addict that although he himself did not use, his girl did and she was 'sick' (withdrawal sickness). This particular plea is effective not only because of the convincingness of its presentation but also because of the values which it invokes; helping another addict when they are sick. A stranger displaying symptoms of withdrawal has a good chance of persuading a dealer to sell to him, but for the purposes of the agent, the degree of sickness required for this would be difficult to feign, and the real sick addict would never make an excuse to 'get off' (inject heroin) elsewhere. As mentioned on page 105, selling to a 'sick' stranger still requires the application of Rule 1a.

Several informants also used the mis-use of certain questions and behaviors as clues to the addict's true identity as an undercover agent. This combined with the application of Rule 1b is very well illustrated by one female addict's account of her encounter with a suspected agent.

> One time this friend A. brought me this guy, B. to buy for. I could smell him a mile off, he was a cop, I'm pretty sure. He asked too many questions, things he didn't need to know as a hype. Hypes don't

waste time with a lot of questions. And he called me up the day before to set up an appointment for the next day. Hypes don't make appointments for the next day; when they need dope, they need it right away. He said he had a lot of people counting on him, their line had gone dry and he needed time to get more money. He said $1, 400. I told him he didn't need that much. If he was a hype, $50 would be enough, at least for the time being. He wouldn't wait until he had more. And if he had a lot of people hung up, they wouldn't wait around until tomorrow either. The kinds of questions he asked were like, "How many people does your man sell to, where's his stuff come from?" that sort of thing.

So I decided not to sell to him directly, because I thought he was a cop. I just wanted to take all the money from him, burn him for trying to trap me. But I did sell to him indirectly, through A., because I needed the dope. I was sick and so I took the chance. I went and got the dope and I must have taken ¾ of every bag for myself. He bought ten $50 bags. I handed it to A., and then A. gave it to B. B. held up the bags, which were practically empty, and said, "Is this stuff as good as you say it is? Oh wow, it looks great!" Then I knew he was a cop, because he didn't even know how much was taken out. Another thing he did which was wrong, he told me to take out my pinch before I brought it back to him. Junkies never tell you that. They tell you not to pinch until you bring it back, and then they promise to turn you on.

The above clues which lead to the addict's intuitive appraisal of whether a stranger is an agent or not have taken the addict quite a distance from the original protective Rule 1, "Never sell to strangers".

DISCUSSION

The preceding section constitutes a summary of a set of behavioral rules which the addict follows in an attempt to avoid arrest for sales of heroin to an undercover agent. This set of behaviors performs the primary purpose of reducing the uncertainty of the outcome of sales to strangers. In a practical sense, by performing such sales according to certain rules, the addict feels he has assured himself of the stranger's identity sufficiently to chance selling to him. Malinowski (1954) discusses this idea of uncertainty reduction in reference to the greater presence of magical ritual under conditions of the greatest uncertainty of outcome. Under conditions where the addict has greater empirical control over the outcome this emphasis on magical ritual is not found.

Thus magic supplies primitive man with a number of ready-made ritual acts and beliefs, with a definite mental and practical technique which serves to bridge over the dangerous gaps in every important pursuit or critical situation. . . . The function of magic is to ritualize man's optimism, to enhance his faith in the victory of hope over fear. Magic expresses the greater value for man of confidence over doubt, of steadfastness over vacillation, of optimism over pessimism (1954:90).

The procedures incorporated into Rule Set 1 are not necessarily magical, although some have magical characteristics, but they do serve the same purpose which Malinowski attributes to magic for primitive man. The entire interaction between the dealer and the stranger involves uncertainty, but where possible or available, this uncertainty is reduced by practical means. This gives the addict greater control over the outcome of the encounter in situations where Rule 1a is used, since he can select not to sell to someone who will not shoot, and know that those who do shoot are not undercover agents. Where Rule 1a, and generally also 1b, is utilized, the procedures contain the requirement for an act which has been defined by other researchers as a ritual — shooting or getting off (Agar 1973, 1975; Howard & Borges 1972). This is a ritual which is secular to us, but according to Agar, it is sacred to many addicts (1975). In the context of Rule 1a, shooting is a required behavior. This use of getting off as a ritual within a larger complex of identity determination emphasizes the sense in which getting off is symbolic of being an addict. In the circumstances discussed previously, this symbolic act is taken as proof of the status of the stranger, and serves to eliminate non-users from the interaction. It thus has the symbolic purpose of establishing identity and the practical purpose of assuring that undesirable identities are not included.

Malinowski also makes a distinction between the application of 'scientific' techniques and those of magic as a means of reducing uncertainty in reference to gardening practices.

Thus there is a clear cut division: there is first the well-known set of conditions, the natural course of growth, as well as the ordinary pests and dangers to be warded off by fencing and weeding. On the other hand is the domain of unaccountable and adverse influences, as well as the great unearned increment of fortunate coincidence. The first conditions are coped with by knowledge and work, the second by magic (1954:29).

For addict behaviors, this distinction is compounded in the application of Rule Set 1. The behaviors characterized as Rules, 1, 1a, 1b, and 1c have in common that they are all dealing with avoiding the same disastrous outcome — being busted. They also involve the same central behavior, sales of heroin to a stranger. Where they differ is in the degree of control which the addict is actually able to exert over the outcome; that is, the extent to which he is able to effectively reduce his chances of being busted. Since Rule 1 is too restrictive, the application of Rule 1a is a rational and empirical response to this difficulty. It is interesting that this Rule is the one which is also most stringent in its requirement of a certain set of behaviors, getting off, which forms a ritual action. Science and ritual in this case are not exclusive, but complementary.

As the addict's ability to control the outcome of the encounter is reduced, because of his failure to require the performance of this ritual, he employs progressively more intuitive, non-empirical measures in an attempt to still achieve that necessary reduction of uncertainty. However, despite the dealer's continued 'confidence over doubt' in the safety of sales to strangers, more busts result from the application of Rules 1b and 1c than 1 and 1a. There is a decrease in the effectiveness of uncertainty reducing measures at really determining the probability of a good outcome from Rule 1 to 1a, 1a to 1b, and from 1b to 1c, and there is a subsequent increase in the subjectiveness of these measures.

WHEN THE RULES ARE BROKEN

Despite the existence of these behavioral rules of protection, addicts do get busted for sales to undercover agents. This appears to happen when Rules 1b and 1c are substituted for 1 or 1a. There is an additional Rule or rationale to account for these substitutions. The critical determining aspect of Rule 1a, permitting it to be as functional as Rule 1 and thus an effective means of control, is the imbedded ritual of shooting. This is included, if at all, only indirectly in Rule 1b, and omitted entirely in Rule 1c. These substitute rules cannot therefore provide the same guarantee of the stranger as Rule 1a. They are inferior behavioral rules when it comes to control of outcome, but they still provide the addict with enough confidence to go through with the transaction.

The justifications which the addict is able to make for his failure to use Rule 1a are invariably given when he has substituted the inferior Rule 1b and/or 1c. His need to offer justifications demonstrates that the central aspect of identity testing, the direct observation of shooting, cannot be omitted without reason. These justifications are the same for failure to follow Rule 1a, and frequently for not using Rule 1 also, and for disregarding the clues of misrepresented identity revealed through the use of Rule 1c. They do not consist of protestations of ignorance or doubts about the efficacy of the Rule Set. Rather, they cite the occurrence of other factors which are given precedence, and which act against rigorously following any of the Rules.

The excuses used by all but one of my informants share the same underlying rationale. The dealer goes ahead with the sale despite his failure to positively ascertain that the stranger is not an undercover agent, and even when he has good reason to believe the stranger is an agent, because of the pressures on him of the needs of his habit. The woman quoted on pages 107 and 108 says, "But I did sell to him indirectly, through A., because I needed the dope. I was sick and so I took the chance". Other justifications are the same type as this, always invoking the threat of withdrawal sickness and the need for money and dope in order to avoid it.

Most addicts, when they are having difficulty maintaining their habit, will take risks which under more beneficial conditions would not be considered. Thus, when the addict feels that he is sick, or in imminent danger of becoming sick, or even on the edge of such danger, he is more liable to sell to a person about whom he still has some reservations. Such action, under other circumstances irresponsible, is excused by his need to avoid withdrawal. Whether or not such conditions were actually present at the time my informants sold to strangers, it is significant that they were claimed. The strongest justifications which the addict possesses are the needs of his habit, and he can use these needs both to explain why he took certain actions and why he omitted certain others.

On his part, the stranger can usually give several reasons why he will not shoot when requested to do so. These reasons are all valid under certain circumstances, and could, if conditions were appropriate, be a legitimate excuse. The most common excuse used by an undercover agent is that he does not shoot, but 'snorts' (inhales through the nose). This also serves to explain his absence of 'tracks' (needle marks), the presence of which is a clue to identifying another addict. An extension of this excuse, given to explain why he cannot snort in front of the dealer, is that he has no time because he has people waiting to buy. This same rationale can also be used when the stranger claims to shoot, but must leave because people are waiting. Another excuse is that they do not use at all, only sell, and finally, that they do not use but are trying to buy for a sick friend.

There are undoubtedly other excuses which could also be interpreted as reasonable. All of these have in common the fact that there really are people whom the addict knows, or knows of, who fit the characterizations given above. There are addicts who snort, although they are in the minority, and non-users who sell, and people who will try to buy for a sick friend. Perhaps under normal circumstances the dealer would not accept any of these excuses as legitimate by themselves. The stranger must present other demonstrations of his trustworthiness, or he will not be sold to. But it is clear that the conditions under which addicts have sold to undercover agents are rarely normal circumstances. The threat or presence of sickness makes the entire encounter an extraordinary one for the addict, irregardless of how frequently he experiences such sickness.

There are several ways in which the addict's actions in regard to sales to a stranger are analogous to a magical rite. The alleviation of uncertainty is only one aspect of this, and it has already been mentioned how the mystical aspect of addicts' actions under Rule Set 1 increases as their control over the outcome decreases. Philip Newman (1974), in a discussion of magic and religion in New Guinea, makes three conclusions about why the practitioners of magic continue to believe in the efficacy of their practices despite many failures.

> In the first place, for people such as these, there is no alternative. Secondly, for the believer in the efficacy of magic, the occasional chance successes are more significant than repeated failure. Finally, explanations for failure are always at hand (1974:324).

That the addict's behaviors under Rule Set 1 have a greater incidence of effectiveness, at least in the case of Rules 1 and 1a, than the magical practices to which Newman is referring, and that they are in fact responses designed to have some effect on actual dangers, does not eliminate the comparisons which can be made using Newman s three points. These become especially applicable in those cases where the requirements of Rule 1a are not met. Whatever the addict does to reassure himself as to the identity of the stranger, he uses the only methods which are available to him. The most reliable method is to see the stranger shoot, but this may not be possible, or eliminate too many potential customers for the needy addict. Instead, the dealer substitutes those alternatives which are possible, relying on his belief that he can trust other addicts, and that he can distinguish real addict behavior from that pretended by an undercover agent. If he elects to make the sale, and he generally has strong reasons for making that choice, then he is necessarily limited to whatever procedures are available.

I have little information on how often addicts are correct in deciding to trust and thus sell to strangers, but the persistence of their confidence in being able to identify an agent suggests that they must be able to do so fairly frequently. Whether they act on this identification depends upon how near to being sick they are. However, that they are sometimes able to use these methods with success makes them feel that they work. In actuality, the intuitive methods used under Rule 1c do 'work' in the sense that they are accurate indicators of how an addict should behave and what he should talk about. In the same way, Rule 1b is an accurate assessment of what other addicts should be, that is, be trustworthy. The failing of these indicators is in the dealer's, and thus the subculture's, inability to provide tests which can reveal the individual who mere ly seems to be trustworthy, or merely adopts these indicators so as to appear to be an addict. The test of shooting works to distinguish between the user and the non-user, but it has no substitute which is as reliable.

The addict's belief in the efficacy of his methods to reveal an undercover agent persists despite his own and other's experiences to the contrary. The failures are less significant, and the major way they are made less significant is through the dealer's ability to provide a reasonable explanation of why they occurred. This explanation places blame not on himself, nor on the system of rules, but on some other factor, such as sickness or need, which clouded his judgement. The dealer's confidence in his methods is fortified and supported by the ready availability of explanations for failures. Seemingly, no addict gets busted except through carelessness resulting from the pressures of his habit.

Thus a bust for sales in no way seems to shake an addict's faith in his ability to control the presentation of his dealing. All nine of the individuals who were busted claimed to have been suspicious of the stranger/agent. None would admit not recognizing that there was something wrong with the stranger's presentation of identity. Perhaps this is just after-the-fact rationalizing but it demonstrates the need of the dealer to always maintain the semblance of control over his fate.

The addict believes and acts in accordance with these beliefs. When the outcome is not what was expected or desired, the beliefs do not necessarily undergo change. There is no need for the beliefs to change as long as there is an explanation for the failure separate from the basic tenets. Such explanations are widely available and are, in fact, built into the system. As an addict, there is always an out in withdrawal sickness.

CONCLUSIONS

This paper has presented a description of a system of rules through which one segment of addict behavior is structured. As such, it is an ethnography of this one small but important aspect of the heroin addict subculture. Rule Set 1 has four rules intended to resolve the problem of potentially being busted for sales to an undercover agent. All of these may be, and with the exception of 1c always are, mutually exclusive operating procedures.

Rule Set 1:

- Rule 1: Never sell to strangers.
- Rule 1a: Never sell to strangers unless you have seen them use.
- Rule 1b: Never sell to strangers unless someone you trust vouches for them.
- Rule 1c: Intuitive methods: identification of the stranger through clues of dress, talk, and behavior.

This rule system changes greatly in ability to accurately predict outcome of the encounter between a stranger and a dealer as one moves down it from Rule 1 to Rule 1c. In the course of this downward movement, the rules employed become less empirical and more intuitive. At the same time, they become less restrictive in terms of who the dealer may sell to.

Rule 1 bans all sales to all strangers, thus severely restricting the dealer's potential set of customers. It is, however, a very efficient and low risk method of avoiding sales to undercover agents.

Rule 1a increases the number of people to whom the dealer may sell by including a test by which the addict is differentiated from the non-addict. It is thus more adaptive to the subculture's reality, and more pragmatic, than Rule 1, although all sales to non-addicts are still excluded. Like Rule 1, 1a is an efficient method of reducing the uncertainty of the encounter's outcome.

Rule 1b further expands the number of individuals to whom the dealer may sell by permitting the possibility of sales to legitimate non-addicts as well as vouched-for addicts. It does, however, require the presence in the interaction of a thirty party besides just the dealer and the stranger in order to be utilized. Because of the need for trust in someone other than oneself, the application of this rule is more intuitive than those above it. And, further, because 'trusted' others sometimes turn out to be informers, it is less satisfactory also in avoiding a bust.

When used as a supplement to 1a and 1b, Rule 1c serves to provide the dealer with an additional measure of the accuracy of these methods and an extra margin of safety insofar as it is effective. When used by itself, Rule 1c permits the greatest number of sales to strangers since it requires no ritual actions, includes both addicts and non-addicts, and eliminates the need for the additional person for verification. This rule, unlike those preceding it, is entirely intuitive and subjective. Although the clues employed by the addict are ultimately based on observed distinctions between agents and addicts, informants were unable to clearly articulate these distinctions as rules for behavior, and their use of such clues was both ambiguous and unsystematic.

The procedures contained within Rule Set 1 are thus seen to cover the entire range of methods of control from those which are 'scientific' to those which are intuitive. Despite the variation between these methods of identifying and resolving the problem of how to respond to a stranger, the entire Rule Set has certain factors in common.

First, in any encounter with a stranger who wants to buy heroin, the dealer will always employ at least one of these rules. Their application is thus very regular and dependable. Second, the addict believes in the efficacy of each of these rules as a means of acquiring control over predicting the outcome of the interaction. Third, when an addict fails to employ the most reliable of these rules, 1 and 1a, he must justify his failure, and the excuses he uses are built into the system and do not cast doubt on the efficacy of the rules themselves. Finally, all of Rule Set 1 serves the central purpose of reducing uncertainty about an encounter which is potentially dangerous. Because of the regularity of occurrence, the faith in efficacy, and the ready excuses for failures, the addict is able to continue to believe that he is indeed behaving rationally and carefully in his interactions with strangers, reducing the uncertainty of the outcome, and thus he need not be unduly anxious about being busted for sales to an agent.

BIBLIOGRAPHY

Agar, Michael, 1973, Ripping and running. New York: Seminar Press.

Agar, Michael, 1975, Into that whole ritual thing: ritualistic drug use among urban American heroin addicts (manuscript).

Howard, Jan & Phillip Borges, 1972, Needle sharing in the Haight. In David Smith & George Gay (eds.), "It's so good, don't even try it once." Englewood Cliffs: Prentice-Hall, Inc.

Malinowski, Bronislaw, 1954, Magic, science and religion. New York: Doubleday.

Newman, Philip, 1974, When technology fails: magic and religion in New Guinea. In James Spradely & David McCurdy (eds.), Conformity and conflict. Boston: Little, Brown & Co.

Stephens, Robert & Stephen Levine, 1971, The 'street addict role': implications for treatment. Psychiatry 34: 351-357.

J. BRYAN PAGE

11

THE STUDY OF SAN JOSÉ, COSTA RICA, STREET CULTURE: CODES AND COMMUNICATION IN LOWER-CLASS SOCIETY

As one of three anthropologists who comprised the socio-cultural component of a trans-disciplinary investigative team for the study of chronic marihuana use and its effects on an urban Costa Rican population,[1] I was assigned to the task of finding long-term marihuana users to participate in the various phases of this study. It had been established during a feasibility study for this research that a population of marihuana smokers with ten or more years of experience with that drug existed in the San José area in sufficient numbers to make such a study possible. We had chosen to work with working-class users and controls, because the confidentiality of the individual subjects' status as a smoker would be insured more easily.

San José, Costa Rica is the largest city in a central cluster of cities which house one-fourth of that country's population. It lies in the middle of a densely-populated intramontane valley whose lands are ideally suited for growing coffee, the principal cash crop of Costa Rica. Compared to other capital cities of small Latin American countries, San José gives the impression of relative prosperity. The visibility of North American chain businesses, such as McDonald's, Kentucky Fried Chicken, Hertz, Holiday Inn and Pizza Hut testify to a strong orientation toward consumer goods and services. The people of San José, even the poorest, are better-dressed than their counterparts in Mexico, or neighboring Nicaragua. The city is dense and bustling with vigorous activity, especially in its center, which contains the main shopping and market areas. In this fast-moving atmosphere where there is money to spare, the street people of San José live and thrive. Often the products of unliveable home situations, these people earn their livings by making something out of nothing. Shining shoes, guarding parked cars, begging, swindling, and sometimes stealing or selling marihuana are some of the means used by the street people to stay alive. The street people, products of years of living by their wits, were our first informants.

Many anthropologists find that their first, and sometimes their only informants are individuals who are somehow set apart from their own societies. These individuals may be leaders or innovators in their own cultures, or they may even be considered deviants and abnormal by their own cultural compatriots. We found in our first contacts with Costa Rican marihuana users that the latter was true. The group of shoeshine 'boys' (the oldest was 49) which formed our first major contact with the marihuana using population is considered quite deviant by the general public in San José. Their mercurial, unattached lifestyles and their histories of frequent arrests set them apart from their fellow Costa Ricans. As we shall see later, these shoeshine 'boys' actively set themselves apart from 'decent' Costa Ricans by means of their linguistic behavior, designed to keep out those who do not share their street-wise way of life.

A field researcher faces several fundamental problems when he attempts to make entry into a social ambience where the participants are involved in illegal activity. Most of these are communication problems. Anthropologists in the field usually confront difficulties in communication because the culture they are studying uses a language that is different from the anthropologists' native tongue. The field team for the study of long-term chronic marihuana use in Costa Rica began its term in the field with what appeared to be an advantage in their already-established familiarity with the Spanish language. We soon discovered that the people with whom we were working spoke a language that was not intelligible to us. This language which perplexed the field team for several months was pachuco, the language of San José street life.

Costa Rican Spanish itself is relatively easy for the foreign Spanish speaker to understand, once he becomes accustomed to stylistic variations in verb person usage, and some relatively straightforward lexical twists. The formal 'Usted' form is used often in situations of some familiarity, and therefore it appears perhaps more often in Costa Rican speech than in that of other Latin American countries. Mothers address small children with the 'Usted' form, denoting social distance that may be eitherdominant to subordinate or subordinate to dominant. Urban Costa Ricans who are on familiar terms with each other use a form similar to the Argentine familiar form, with an accented last syllable. Rural Costa Ricans use the 'Usted' form almost exclusively. Among the lexical adjustments in Costa Rican Spanish, one of the most striking is the use of 'macho'. Costa Rican Spanish is apparently unique among Latin American dialects in using 'macho' to point out skin, hair, or eye coloring instead of manliness or maleness. Costa Ricans call each other 'macho' or 'macha' if they happen to have blue eyes or light skin or blonde hair.

These minor differences in Costa Rican Spanish presented no real difficulties for the field team, because adjustment to them progressed at roughly the same rate of adjustment to using Spanish after a period of exclusive English use. As we established early informant contacts we found that when they spoke directly with us, we understood fairly well, but when the same people spoke among themselves, we could not understand at all. These individuals were using a code of verbal communication which was designed to keep outsiders of doubtful confidence from understanding important segments of communication. As foreigners who at times were even suspected of being special police agents, we found that although many of our early informants were often very cooperative, they persisted in speaking a language that we could not understand. By their use of pachuco in its most elaborate and alien forms, the informants were able to hide the significance of what was taking place during the course of everyday activity, thereby frustrating our investigative efforts.

Typical of the difficulties we experienced in the early stages of investigation is the following scenario: While seated with one of the shoeshine 'boys' and having what I consider a productive and enlightening conversation on the subjective effects of marihuana, three other shoeshine 'boys' approach my informant. The four immediately launch into a rapid and bewildering discussion in which I am able to pick out only isolated words. They occasionally make reference to me which I am also unable to decipher. When his companions have left, my informant returns to me, and excuses himself, explaining that they had had some important 'business' to discuss. When I press him for a translation of that verbal exchange, he changes the subject.

Breakthroughs in our problems with communication did not begin to happen until four months after the study had begun. By that time, we had established rapport with a few crucial individuals who decided that we were not agents of the police. They began explaining some of the lexical changes which differentiate standard Costa Rican Spanish from pachuco. Most of these are known to average Costa Ricans who are not necessarily participants in San José street culture and they are not crucial terms for understanding conversations about illegal or clandestine activities. Such lexical variants include several articles of clothing and some anatomical terms. The first Pachuco word explained by informants was 'cruz' (lit: cross) which means 'shirt' in that argot. 'Cruz' was followed by 'caballo' (lit: horse) which means 'pants' and 'cacho' (lit: horn) which means 'shoe'. Most Costa Ricans could tell you the meanings of these words and others, such as 'güachos' (eyes) and 'piano' (teeth), even though 'decent' Costa Ricans hardly ever incorporate this kind of vocabulary into their own conversations.

The addition of the first pachuco words to the field term's repertoire was not immediately helpful in making sense out of conversations among our informants, because the bulk of these conversations contained lexical variations and other codes which had not been revealed. We felt that the informants' confidence in the field team was growing when some more crucial lexical items were translated into Spanish for us. Because of our open and often-repeated interest in marihuana, some informants saw fit to reveal some of the elementary terms for the drug itself:

mota-marihuana, meaning the material itself
moto-marihuana cigarette
grifa-marihuana (also material)
grifo-adjective meaning 'stoned', or under the influence of marihuana
zoncha-marihuana
monte-marihuana
cochinada-marihuana
manteca-marihuana
mariquita-marihuana
La que traba'-marihuana
marilú-marihuana

We quickly included these terms in our own vocabularies, and found that our indication of interest in this colorful style of speech led to further revelations of terms that were more central to the business of distribution and sale of marihuana in San José. These contributions were accompanied by other terms which are important to the pursuit of business-as-usual on the streets of San José.

The sale and distribution of marihuana in San José is described by Pachuco terms with a distinct medical flavor. Vendors of marihuana are commonly called 'médicos' (lit: medical doctor) or 'doctores'. The verb used to express the act of selling marihuana is 'recetar' (lit: to give a prescription), and the vendor's customers are called his 'patients'.

The single most crucial and most difficult to decode pachuco word involving marihuana traffic was 'galeta', which has no literal Spanish meaning. 'Galeta' is used to designate both the storage place of a cache of marihuana and the runners employed by vendors to make deliveries and to serve individual 'pacientes' (patients). 'Galeta' was never really explained to any of the field team, but we managed to decipher it through its contexts of use on the streetcorner, both in association with other marihuana-related pachuco words and in association with marihuana selling activities. When a dealer said, "Vaya a la galeta y saque cinco motos bien gruesos,"[2] we could discern one of the meanings of this word. However, when he then said, "Vaya al bar y búsqueme a mi galeta; dicen que anda por ahi,"[3] we were once again confused, because from all indications, what we had thought to be stationary and inanimate had become animate, and probably human. Finally, after continued repetition of these and similar usages of the word, we were able to ascertain that it carries both meanings. It should be noted that the informants never encouraged us to use this word. When we began to put 'galeta' into our conversation, it was not received by the informants with the same enthusiasm as many of the more elementary Pachuco terms. Perhaps the informants were still ambivalent about giving the field team a real key to understand conversations about illegal activity.

Illegal streetside activities in San José include minor gambling fraud and petty thievery, and these pursuits also have a terminology couched in Pachuco. To steal is to 'pegar un descuido' (lit: to catch off-guard), and stolen articles are called 'descuidos'. The shell game used to swindle unsuspecting victims, called 'chapas' after the bottle caps which replace the shells, employs a 'gancho' (lit: hook) who operates as the 'shill' or 'plant' to convince the victim that he has more than an even chance to win.

Such a terminology system may seem to be trade-specific, as many trades have complex specialized vocabularies. For example, Kochman (1972: 243) cites the pimp style of speech and vocabulary as a necessary tool for this particular street survival tactic among urban blacks in Chicago. Theft and gambling related items, on the other hand, are apparently known to all pachuco speakers, which may be indicative of a lack of specialization in illegal activities. The most fluent and colorful speakers of pachuco are diversified in their economic activities, and they do not concentrate on any single illegal activity. Using the typology of marihuana users described in an earlier work we call these people street-movers (Page 1974). The street-movers typify a style of economic activity where wage-earning labor is not usually included in the repertoire of strategies for making a living. He may work his 'chapa' swindle in the streets or sell marihuana or resell 'descuidos' for a profit. He has learned during his years of experience of making something out of nothing that this diversification is both adaptive and necessary. Thus, the true pachuco speaker has a wide range of specialized vocabularies for illegal activities because he has participated at one time or another in all of these activities. Such a set of vocabularies is not common knowledge to the general Costa Rican population as with the pachuco parts of the body. Those who did not grow up with the vocabulary of illegal activities cannot usually participate in them.

Even so, lexical changes alone cannot make pachuco unintelligible to outsiders for very long. Since the purpose of a street language like Pachuco is to exclude individuals who come from a different segment of the same society as the speakers of pachuco, it must keep ahead of society's general adoption of pachuco lexicon by constantly adding and modifying. The adoption of a subculture's colorful expressions by the society at large is described for the case of black street talk by Brown (1972:136-137). He cites the now general usage of the expressions 'sump'n else', 'cop', 'boss', 'strung-out' and 'uptight' as examples of this pattern of adoption.

The pachuco defense for such adoption practices is illustrated in the following example: During the field team's stay in Costa Rica, the police department, called the civil guard, received new black-and-white patrol cars which replaced older red-and-black models. Within two weeks, we noticed that one of our most creatively pachuco-speaking informants was calling rice and beans 'patrulla', or 'patrol' because the dish has the same colors as the patrol cars. This expression later appeared in other informants' speech. Pachuco texts of the early 1960's appeared in one of San José's newspapers, and basic vocabulary items in the pachuco of that time are different from those in current use. For example, the word 'feo' (meaning 'ugly') is very important in Costa Rican conversation, and the pachuco equivalents are 'gacho' and 'güeso'. In 1962, the pachuco words for 'ugly' were 'pelis' or 'furris'. The rate of turnover for pachuco lexicon must be comparatively rapid if basic vocabulary changes take place within ten years.

More successful at preventing the outsider's participation in pachuco verbal interaction are the word games. Practiced pachuco speakers are so adept at these games that they perplex all but the most practiced pachuco speakers. Most prominent among these word games is the art of talking 'al veres' (al revés, or 'backwards'). The pachuco speaker who is talking 'backwards' reverses the syllabic order of two-syllable words, and places the last syllable first on most words of more than two syllables. Thus, the phrase 'saque el moto' (get rid of the joint) becomes 'quesa el tomo'. Single syllable words are usually left as they are in 'backwards' talk, except 'sí' or 'yes' which becomes 'is'. The utterance 'sí, chavalo', (that's right, clown) then, becomes 'is, valocha'. We have observed some of our informants perform long utterances in which each word consistently followed the 'al veres' rules. These utterances were unrehearsed and spontaneous, and, most amazingly, the Pachuco-speaking companions in these cases understood the entire utterance, and acted according to instructions contained in that utterance.

The fluency of some of the pachuco-speaking informants in talking 'backwards' leads us to believe that there are adaptive rewards for the street people who develop this skill. Those who are most likely to participate in illegal economic activity seem to be the ones who have the talking 'backwards' skill most highly developed. Perhaps the addition of the word game dimension to the lexical and stylistic characteristics of pachuco speech allows for added protection from prying ears in situations where a conversation overheard and understood can be incriminating. Some pachuco speakers who are very fluent in their own way do not pepper their speech with 'backwards' utterances nearly as much as others. These individuals do not tend to be engaged in illegal economic activity, so their pachuco usage serves as a symbol of solidarity with other pachuco speakers and as a means of keeping a relatively small area of their activities confidential. Some informants are using pachuco word games to keep the police from discovering that they sell marihuana or play 'chapas', (the illegal shell-game mentioned earlier) while others use the pachuco style and lexicon to keep their family from finding out when and where they are going to smoke marihuana, or have extramarital sex.

Pachuco stylistic variations add color to the other characteristic patterns that keep it out of the average Costa Rican's reach. Currently, it is popular to address other people as 'maje', which literally means 'fool', but because of the frequency of its use 'maje' has come to mean the same as the swingy use of 'man' among hip English speakers. Pachuco speakers also attempt to maximize the deep resonances in the voice when they speak to add bravura to their speech.

None of the field team in the Costa Rica study actually became fluent in pachuco. This would have required more development time than was available. Nevertheless, we managed to break the pachuco codes sufficiently to understand conversations, thereby picking up crucial bits of communication. Constant exposure to the variations of pachuco speech was one of the most fascinating aspects of the field team's research experiences. An important product of this exposure is a compiled pachuco lexicon in which there are already over 1500 items. Certain pachuco speakers were able to provide conversations which added a clearly aesthetic dimension to our experience with this colorful street language. The following piece is extracted from interview materials of an informant whom we considered to be the 'Shakespeare' of pachuco:

Pachuco

De cabrillo, bueno lo que uno más se puede acordar es de las varas de la escuela y pajas, después de lo que uno ha sufrido, y todo esa vara ... bueno son cosas que en el plan, son gachas recordarlas, entiende? El plan es güeso, pero diay, es tuanis también, porque diay, uno remueve las varas de chamaquillo. Son varas que uno no platica con nadie, que se van al olvido, verdá, pero el plan tuanis. Esa vara es como cuando un agüelo agarra al nieto y se pone a contarle la vida de él, estando joven. El plan es, bueno digamos de sufrir en el sentido de, cómo le explicara yo? Bueno en parte tenía culpa de que sufriera yo mis padres, verdá?

Polite Spanish

De nino, lo que uno más se puede acordar es de los asuntos de escuela y cosas, después de lo que uno ha sufrido, y todas esas cosas ... bueno de todos modos, son feas recordarlas, entiende? La idea es fea, pero es linda también, porque uno remueve cuestiones de la ninez. Son cosas que uno no platica con nadie, que se van al olvido, verdad, pero la idea es linda. Esa idea es como cuando un abuelo pone el nieto en su rodilla y se pone a contarle la vida de él, estando joven. La idea es, bueno digamos de sufrir en el sentido de, cómo explicara yo? Bueno en parte tenía culpa de que sufriera yo mis padres, verdad?

The English translation of this segment would read more or less like this:

> As a little kid, well what you remember most are things about school and jive, after what you've suffered and all that stuff ... well they're things that anyhow are shitty to remember, understand? The idea's ugly, but hell, it's groovy too, because you remove the shit of childhood. It's stuff you don't talk to anybody about, that go into forget-

fulness, right? But the idea's groovy. That bullshit is like when a gram'pa grabs his grandson and gets to telling him about his life as a young man. The idea is well let's say to suffer in the sense that, how should I say it? Well, partly my parents were to blame that I suffered, right?

NOTES

1. A research contract funded by the National Institute for Drug Abuse (Contract # N01-Mh3-0233-ND-).
2. Go to the marihuana cache and get out five very thick cigarettes.
3. Go to the bar and look for my runner for me; they say he's around there.

BIBLIOGRAPHY

Brown, Claude, 1972, The language of soul. In Thomas Kochman (ed.), Rappin' and stylin' out: communication in urban Black America. Urbana: University of Illinois Press, pp. 134-139.

Kochman, Thomas, 1972, Toward an ethnography of Black American speech behavior. In Thomas Kochman (ed.), Rappin' and stylin' out: communication in urban Black America. Urbana: University of Illinois Press, pp. 241-264.

Page, John Bryan, 1974, Marihuana use in San José, Costa Rica from the perspective of user types: a new approach to smoking set and setting. Paper presented at the American Anthropological Association meetings in Mexico City, November, 1974.

HALLUCINOGENS AND SENSORY STIMULATION

One of the basic characteristics of drug use by traditional peoples, was the use of naturally occurring psychotropic substances. These leaves, roots, fruits or whole plants were taken to induce visual, mental, or tactile hallucinations. Thorough documentation exists concerning the use of kava in Oceania (Marshall, 1974), tobacco (Wilbert, 1972, 1975) and peyote (La Barre, 1960, 1969) among American Indians, yagé (Harner, 1973) and ayahuasca (Reichel-Dolmatoff, 1969) in South America, cannabis in Africa (du Toit, 1974, 1975), Amanita muscaria among Norsemen, and such plants as henbane, mandrake and thorn apple among early European witches.

This section contains two chapters, one on the Tsonga of southeast Africa, the other on the traditional settlers in the New World. The first suggests auditory hallucinations as a result of drug use, the second that drug induced visual and mental hallucinations resulted in out-of-body experiences.

BIBLIOGRAPHY

du Toit, Brian M., 1974, Cannabis sativa in Sub-Saharan Africa. South African Journal of Science, 70.

du Toit, Brian M., 1975, Dagga: the history and ethnographic setting of Cannabis sativa in southern Africa. In Vera Rubin (ed.), Cannabis and culture. The Hague: Mouton Publishers.

Harner, Michael J., 1973, Common themes in South American Indian yagé experiences. In Michael J. Harner (ed.), Hallucinogens and Shamanism. New York: Oxford University Press.

La Barre, Weston, 1960, Twenty years of peyote studies. Current Anthropology.

La Barre, Weston, 1969, The peyote cult (enlarged edition). New York: Schocken Books.

Marshall, Mac, 1974, Research bibliography of alcohol and kava studies in Oceania. Micronesica, 10(2).

Reichel-Dolmatoff, Gerardo, 1969, El contexto cultural de un alucinógeno

aborigen: Banisteriopsis caapi. Revista de la Academia Colombiana de Ciencias Exactas, Fisicas y Naturales, XIII.
Wilbert, Johannes, 1972, Tobacco and Shamanistic ecstasy among the Warao Indians of Venezuela. In Peter T. Furst (ed.), Flesh of the Gods. New York: Praeger Publishers.

THOMAS F. JOHNSTON

12

AUDITORY DRIVING, HALLUCINOGENS, AND MUSIC-COLOR SYNESTHESIA IN TSONGA RITUAL

The Tsonga live on either side of the Mozambique — South African border, and thus are of major contemporary interest as a vital factor in the policy of detente between the newly de-colonialized nation to the east and the bastion of apartheid to the west. Often referred to in the literature as the Thonga or Tonga, the two million or so Tsonga have recently adopted the name Shangana-Tsonga in order to be clearly distinguished from the well-known groups Plateau Tonga, Zambesi Tonga, and Inhambane Tonga, from whom they are culturally and linguistically distinct. Shangana derives from the name of an ancient Tsonga chieftain, Soshangane. The Tsonga are a patrilineal, virilocal Bantu-speaking people, a considerable number of whom propitiate their ancestor spirits and engage in polygyny. In addition to the staple maize, the Tsonga plant squash, pumpkins, sugarcane, and groundnuts. A few cattle, goats, sheep, and fowl are kept. Male migrant labor is now a key factor in Tsonga economic life.

The Tsonga possess two social institutions in which rhythmic sensory stimulation via fast drumming, and the ingestion of the hallucinogenic plant-drug Datura fastuosa, are prime mechanisms for the achievement of an altered state of consciousness.

1. 'Khomba', the Tsonga girls' initiation rites, are a series of structured fertility rites held every year after the May harvest. Dance, mime, song, drumming, and drug-ingestion are the basis for an integrated set of symbolic activities.
2. 'Mancomane', the Tsonga exorcism rites held mainly during the period January-March (the South African summer), is a series of healing and transformational rites involving the expelling of undesirable alien spirits from possessed individuals, and the graduation of fledgling exorcists.

In one of them — 'khomba' — occurs the interesting phenomenon music-color synesthesia.

Each of these social institutions comprises a sequence of symbolic activities bearing mystical titles. This indigenous classification of the components of the rites facilitates the approach, comprehension, and participation of the novices who pass through them as a group. Edmund R. Leach, in his classic essay on ritual discusses the Lévi-Strauss viewpoint that "the drama of ritual breaks up the continuum of visual experience into sets of categories with distinguishable names and thereby provides us with a conceptual apparatus for intellectual operations at an abstract and metaphysical level" (1968: 524). For the Tsonga, who place high value upon musical performance in ritual contexts, the continuum includes aural experience as well as visual.

Each of the two social institutions named features a transformational process — from girlhood to womanhood in the one and from possessed person to practising exorcist in the other — during which there occurs an anticipatory build-up toward a climactic point in the rites. The stages of the build-up contain a vast number of symbols which depend for their cultural and contextual interpretation upon the attitudes and perceptions of the novices, hence the strict authoritarian control and direction by exegetes. This can also be viewed in reverse: the musical, social, and ritual ambience of the scene of the rites is such that tribal authority and gerontocracy is maintained and reinforced. The changed status of the group of individuals passing through the rites in no way challenges the hierarchical order; it is but one revolution in the Tsonga calendric cycle of ritual events. E. D. Chapple & G. S. Coon refer to such rites as rites of intensification: "A rite of intensification conditions the people to the new relations to follow by building up interaction in habitual channels to a high pitch of intensity, through the use of a wealth of symbolism Each symbol used refers to the context of situation of the interaction of the celebrants in terms of their technology. The reason for this is that the mechanism of the conditioned response, through which the symbols obtain their meaning, depends on a regularity of repetition" (1942: 528).

The girls' initiation rites and the exorcism rites share certain interesting features. Most important among the observed commonalities is the controlled use of Datura fastuosa ('mondzo', or 'muri wa ku bonisa' — 'that which opens one's eyes'), a huallucinogenic subspontaneous plant of the Solanaceae family bearing ovate-oblong leaves, blackish-brown seeds, and purple flowers. Its roots are ground and brewed to make a thick tea which, when administered by an officiant in reasonably small doses, produces hallucinations and disorientation in the subject. Other observed commonalities include the following:

a. Loud, fast drumming at approximately the frequency of alpha waves (8-13 c.p.s., the basic human brain wave), probably employed to bring about auditory driving.
b. The use of polyrhythms, probably employed to bring about auditory driving in a group of individuals, each of whom possesses a different basic brain wave frequency.
c. Violent, energetic dancing and other prolonged kinesthetic activity such as miming, bringing about hyperventilation, low blood glucose, and high adrenalin flow.

d. Insistent reiteration of commands and suggestions by a feared authority figure, probably constituting an example of the use of hypersuggestibility.
e. Perceptual hypnotic mechanisms such as the repeated waving (in front of the subject) of a large colored headdress or 'milala' palm-leaf branch.
f. Rhythmic tactile reinforcement in the form of beating with a switch or shaking of a blanket.
g. Prolonged exposure to the rays of the midday sun, probably in an attempt to reduce biochemical resistance to the drug's effects.
h. Immersion in water (the 'crossing-over' component of the rite-of-passage sequence — separation, liminality, and reincorporation).
i. The use of a large sea-shell (i.e. water-associated object) as a ceremonial container for the drug to be administered.
j. The selection of a site near a tree, i.e., virile 'phallic' object containing white sap ('semen', 'mother's milk').
k. The consumption of a small quantity of human fat or powdered bone (in initiation and exorcism this is to obtain protection from witchcraft by doing that which witches do, i.e., eat human flesh; in the trial-by-ordeal it is to force the accused to do in daytime that which he committed at night, thus tricking him into 'losing' his human form, i.e., his normal mental state).

My working hypothesis is that the observed commonalities (a) through (g) exert a cumulative, complementary influence upon each other, and that all of them aid in the reduction of biochemical resistance to the drug's effects. Observed commonalities (h) through (k) represent culture-specific traditional beliefs which shape the motivations, expectations, and attitudes of the participants, maximizing the goal-oriented direction and guidance (by the power-wielding elders) of the participants' visions, supernatural voice-hearing, and other hallucinatory experiences.

Tsonga hallucinatory experience is thus (i) culturally patterned; and (ii) a mechanism for reinforcing the power, prestige, and authority of the tribal elders (as the controllers of drug-use) over the rank and file of the rural community (Figure 1).

As part of the evidence to support the hypothesis, I now present summarized data collected in the field, being my eyewitness account of the ritual use of Datura fastuosa and auditory driving within two Tsonga social institutions: initiation and exorcism.

DRUG-USE IN THE TSONGA GIRLS' INITIATION RITES

'Khomba' the Tsonga girls' initiation school, teaches the women's role of husband-pleaser, infant-bearer, homekeeper, and tiller of the soil,

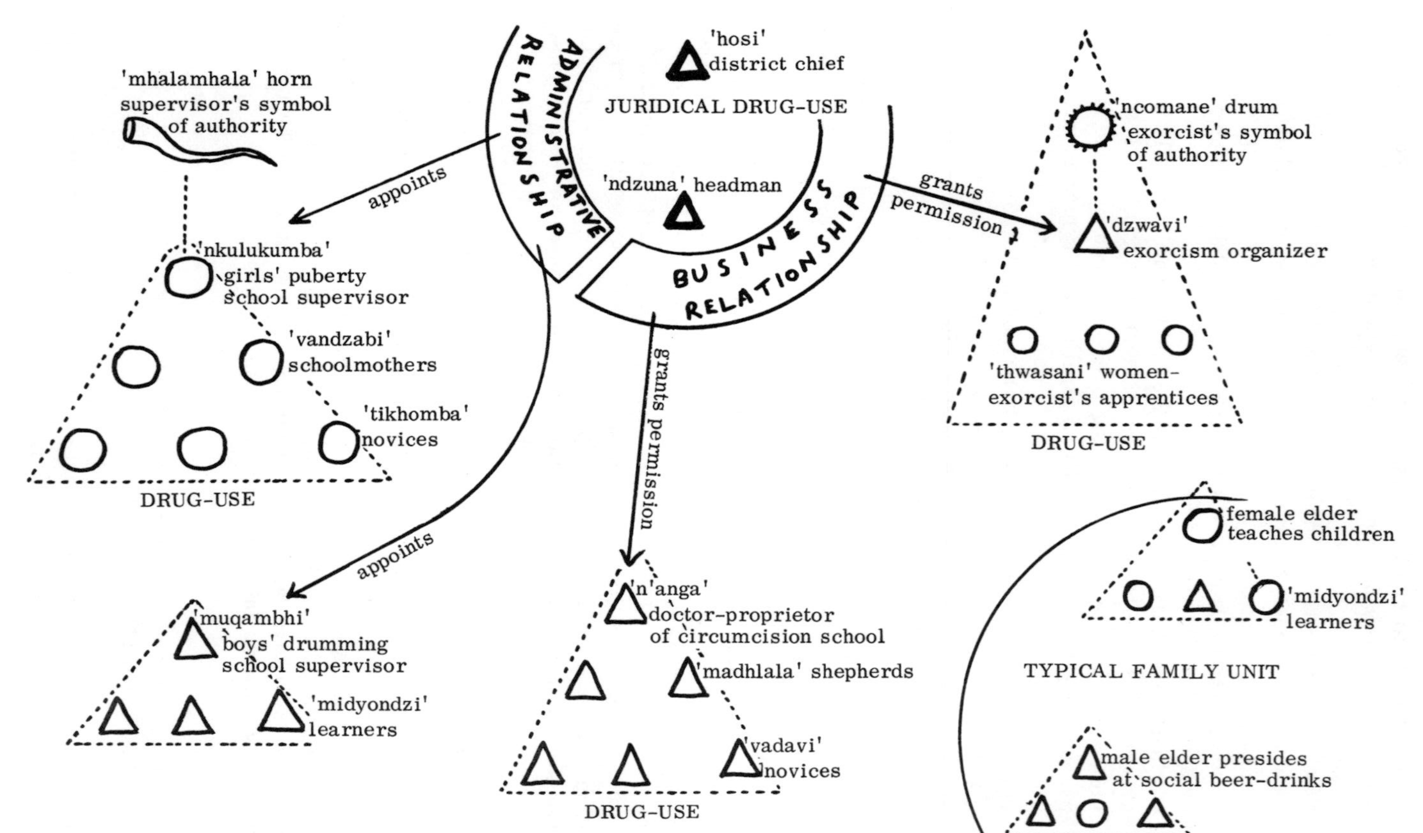

Figure 1. Tsonga chain of ritual authority featuring drug-use control

in that order. It also provides eligibility for marriage, thus bringing cattle to the girl's father. This last is an important economic consideration, for even after graduation through the rites and subsequent marriage, a girl does not fulfil her social obligations until she has born an infant. Should she prove infertile, the cattle will have to be returned and disgrace will follow. 'Khomba' therefore emphasizes fertility, and the drug-use, recitation of secret formulae, dancing, and miming which comprises the initiation rites is largely aimed toward this goal.

Administering of the Datura fastuosa must be preceded by a complex series of mimes and energetic dances, accompanied by loud, fast drumming in polyrhythms, many repeated suggestions from the supervisor, the waving of a large yellow headdress before the eyes, spanking with a switch made from the Datura plant, disrobing and exposure to the sun, immersion in the river, and a strange water-rite in which initiates stretch a skin over a large container of water while offficiants puncture it with poles and swish the water below ('flow of amniotic fluid at parturition'). The series of mimes leading up to the climactic drug-taking ceremony is shown in Figure 2.

During part of the rites girls wear blue-dyed salempores, paint their faces blue, and erect a blue flag. Under the drug they are required to perceive bluish-green patterns; it is more than coincidence that the small snakes (Dendrophis subcarinatus) which inhabit the eaves of Tsonga huts are bluish-green and are considered to embody ancestor-spirits. Called 'xihundze' locally, the snakes are revered and never harmed except by the foolish. The following proverb is commonly heard.

> U nga dlayi nyoka u ndzuluta, ta mincele ta ku vona.
> Swikwembu swa ku vona, swi ta tirihisela.
> Do not whirl a snake on high if you should kill it.
> The spirits will see you and exact revenge.

The bluish-green perceivings of the drugged initiates is thus culturally patterned; the cones, spheres, and other geometrics of the human optical system become amenable to observation under the effects of Datura fastuosa, take on the color prescribed by the culture, and are associated with the fertility gods. The perception of snakes by drugged subjects is not, of course, culture-specific. Masters and Houston report several drug experiments involving the perception of snakes: "dragons and snakes, especially, are at home here and are not fighting" (subject, after an experiment, in Masters & Houston, 1966: 286). But the Tsonga association of blue-green, snakes, and spirit-communication is culture-specific. Furthermore, cultural patterning is to be seen in the Tsonga association of water (purification, cleansing), 'crossing over', adulthood, and fertility (amniotic fluid). The Tsonga term for river-mouth ('nyanzwa') is the same as that for uterus, and numerous terms connected with puberty school musical instruments are related to both fertility and water (Figure 3).

For drug-ingestion the initiates lie on a mat which 'separates them from the dust they knew as children'. They prepare for their 'journey of fantasy' ('rendzo ra miankanyo'), and sing of 'crossing over', as follows:

Call: She is mature, my child
The ship lies on the far bank of the river.
Response: 'Iye, iye', go home (repeat several times).

The shoulder-high mime, with the novice (hoisted high by the others) wielding two sticks to demonstrate labia minora elongation measurements

The tree-climbing mime, with the novice being beaten and a third novice in quasi-foetal position on the ground

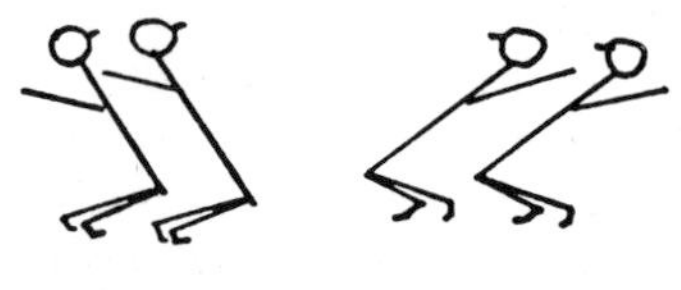

The backward-leaning 'prisoner' dance, with a song about not forgetting one's identification passbook in the town

The mime in standing position with hands on head

The bangle dance, with arm outstretched to receive bangle for protection from barrenness by witch-craft

The hands-on-hips childbirth dance, with a novice in quasi-foetal position

The firewood-gathering mime with stick tapped on the ground

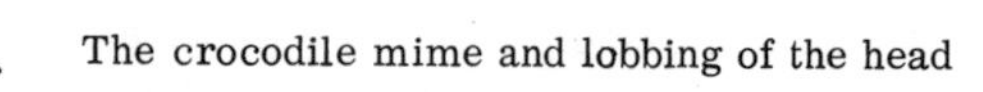

The crocodile mime and lobbing of the head

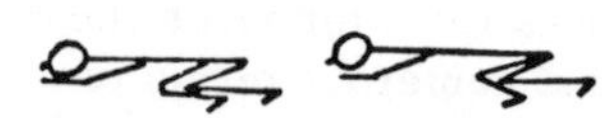

The 'baby-crawl' across the river-bed

Figure 2. The series of mimes leading up to drug-ingestion in the girls' initiation school (note the progressive upward 'growth')

Plate 1. Novices of the Shangana-Tsonga girls' initiation school ('khomba') in the Northern Transvaal. The blue-dyed uniforms and blue paint on the face have religious significance, being associated with the bluish-green snakes called 'xihundze' (Dendrophis subcarinatus) which inhabit the thatched roofs of Tsonga huts, and which are revered as ancestor-spirits.

Plate 2. Prior to a religious rite in which a hallucinogenic plant-drug is administered in order to facilitate supernatural voice-hearing, warm-up dances are performed. This one is called 'nanayila'; novices wave wooden hatchets and blow metal whistles in time to fast drumming by middle-aged women.
Plate 3. The fast, energetic dancing seen here is a mechanism aiding the supernatural voice-hearing goal: it causes low blood glucose, high adrenalin flow, and possibly hyperventilation, thus reducing resistance to the drug's effect.
Plate 4. After the 'nanayila' dance an even more energetic dance ('managa') is performed by two novices at a time. In the background is the hand of the supervisor, who is 'consecrating' the ground with black medicine. The latter brings black rain-clouds, which makes the crops fertile. 'Khomba' is believed to make the girls fertile.

Plate 5. The first of a series of important mimes performed by novices to the rhythm of the drum. Bodies are low, for the 'baby-crawl' mime.
Plate 6. Bodies rise higher for the firewood-gathering mime. The long, twisted antelope horn seen here is used to produce frightening supernatural bellowing sounds in the night, represented as the sound of spirits. It is also used to deflower the novices.
Plate 7. Bodies rise higher for the hands-on-head mime. Note pile of discarded clothing — novices take a new name and fresh clothing after the rites.

Plate 8. The climax of the 'rising posture' series of mimes. Each girl is shoulder-hoisted in order to demonstrate elongation measurements — the extent of achieved elongation of the 'labia minora'.
Plate 9. In preparation for drug-ingestion, a novice lies in quasi-foetal position on the 'milala' palmleaf mat which separates her 'from the dust she knew as a child'. The entire vision sequence may be seen as a period of transition and crossing over, from childhood to fertile womanhood.
Plate 10. The 'doctor' wraps each novice in a blanket and selects a clay square in which straws have been stuck. This is inserted between the legs to represent the re-growth of pubic hair which has been shaved as an act of separation. The bangle before the kneeling woman will be given to the graduating initiate as a charm against sterility by witchcraft.

Plate 11. The 'doctor' brings the drug-potion (a brew from the plant Datura fastuosa) in a ceremonial shell, and carefully administers a measured draught to each novice, who then customarily reports the hearing of supernatural voices and the perception of bluish-green patterns.

The officiant sprays the initiates with saliva; this is called 'ku pela marhe', to 'cross over' water. She produces a large shell containing Datura fastuosa and chants the following:

Solo chant: One digs up the medicinal plants.
Take the medicine of which you have heard so much!

Cocooned in colored blankets, the drugged initiates are beaten with Datura switches and told that they are expected to hear the voice of the

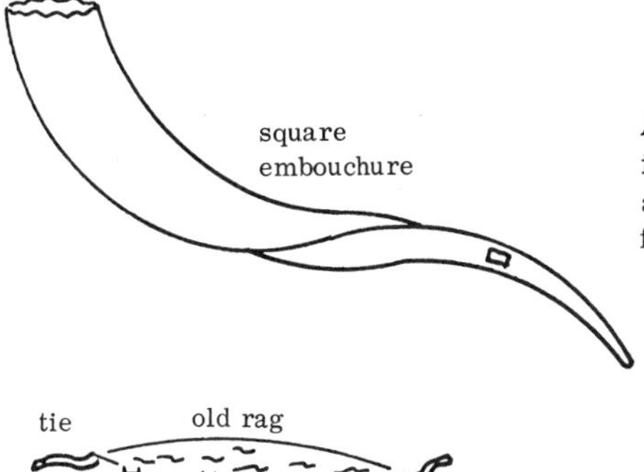

A. The kudu horn, symbol of authority of the puberty school supervisor and means of deflowering initiates, is flushed with water before blowing.

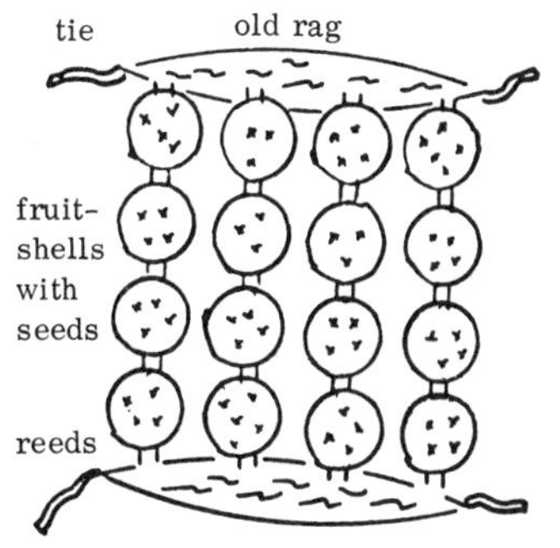

B. The fruitshell leg-rattles, whose sound is 'the voice of the gods' which guides puberty school dance-steps, are also used in Tsonga rain rites.

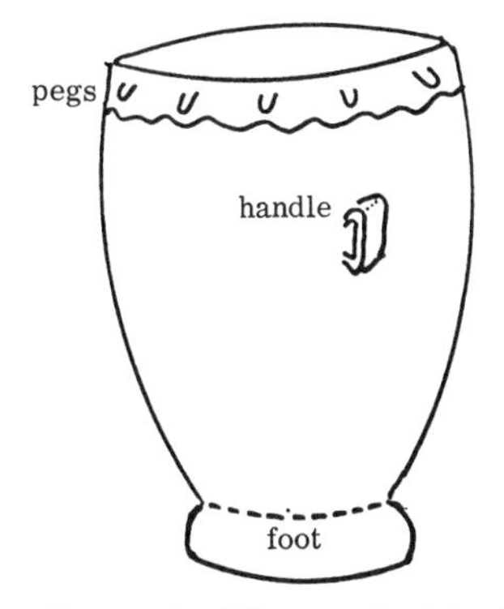

C. The pegs of the puberty school drum are called 'penis', and the hole in the bottom is referred to as 'vagina'. Tension of the head forecasts rain/no rain.

Figure 3. The musical instruments of the girls' puberty school possess fertility symbolism.

fertility spirit. The initiates, in a dream-like state, find spirit-communication easy, for the Tsonga customarily communicate with their gods in dreams (an instance of biochemistry-plus-culture complementarity). Drug-use for spirit-communication has been widely reported in the psychological literature (Blum, 1964:6; Masters & Houston, 1966:257; Barber, 1970:36), and the importance of hypersuggestibility has been emphasized by Tart: "the person comes to rely more on the suggestions of the shaman ... the suggestions of the person endowed with authority tend to be accepted as concrete reality" (1969:17).

The initiates are taught that the mime-specific drumming rhythms produce the bluish-green patterns. This type of music-color synesthesia has been reported in the psychological literature:

> ... in one study (Klee, 1963) the experimenter clapped his hands in the air while the subjects, who had received a high dose of LSD, were observing visual patterns with their eyes closed; the subjects typically reported that they saw flashes of color in time with the clapping. In another study (Guttmann, 1936) mescaline subjects were exposed to music while they were perceiving lines and patterns with their eyes closed; the subjects typically reported that the lines moved and changed colors in harmony with the music. Similarly, a third study (Hartman & Hoolister, 1963) showed that, as compared to non-drug controls, drugged subjects exposed to pure tones reported significantly more colors and patterns elicited by the tones (Barber, 1970:35).

Much separational and phallic symbolism permeates the 'khomba' school, emphasizing the initiates' newly acquired adulthood and possibility for motherhood. For instance, clay squares bearing porcupine-like straws are inserted between the legs of initiates, representing the regrowth of pubic hair which has been shaved off prior to the rites; a tree which is climbed is referred to as the 'xipingwana' (drum-peg); and initiates are required to squat upon an upturned elongated drum.

Infertility, when discovered after the rite, is attributed to witchcraft from enemies; this is true even in the case of literate schoolgirls, showing that Western-style education is not necessarily dissonant with traditional beliefs. To combat sterility from witchcraft the officiant mixes a little human fat or powdered bone with the Datura fastuosa, this being thought to provide an antidote. Fresh Tsonga graves are frequently disturbed during the night by prowling hyenas; the flesh-eating which occurs is attributed to witches, whose activities can be countered by re-enacting the witch-role.

To conclude this account of Tsonga initiation use of Datura fastuosa, it is appropriate to point out certain interesting cross-cultural comparisons. My findings concerning the Tsonga association of water, snakes, and fertility closely match S. G. Lee's findings in his 'Social influences in Zulu dreaming' (1958:265-283), and my findings concerning spirit-communication during initiation drug-use tally closely with the findings from research carried out among the Fang of northwest equatorial Africa (see Balandrier, 1963:226; Fernandez, 1965:902-929; Swiderski, 1965:541-551; and Pope, 1969:174-184).

AUDITORY DRIVING IN THE TSONGA GIRLS' INITIATION RITES

Barber has noted (1970:9) that hallucinogens induce changes in audition. It is also true that certain kinds of audition can intensify the biochemical effects of hallucinogens. In fact, audition and drug-ingestion can be seen as complementary mechanisms for achieving altered states of consciousness, particularly in certain non-Western drumming rituals. In the Tsonga girls' initiation rites, each dance, mime, and symbolic act, including the drug-ingestion act, is accompanied by rite specific drum patterns. These are performed by an exegete loudly and at high speed, close to the heads of the novices, on an untuned drum of general low frequency. The auditory and psychological effect of such a drum is broad, for it is not limited to the single nerve pathway taken by a pure-toned instrument of specific pitch such as a flute. Psychological experimenter Neher states that:

> A single beat of a drum contains many frequencies. Different sound frequencies are transmitted along different nerve pathways in the brain. Therefore, the sound of a drum should stimulate a larger area in the brain than a sound of a single frequency. A drum beat contains many low frequencies. The low frequency receptors of the ear are more resistant to damage than the delicate high frequency receptors and can withstand higher amplitudes of sound before pain is felt. Therefore, it should be possible to transmit more energy to the brain with a drum than with a stimulus of higher frequency ... The range of individual differences in basic brain wave frequency is from around eight to thirteen cycles per second ... We expect to find, therefore, a predominance of drum rhythms in the range of slightly below 8 to 13 c.p.s. in ceremonies that precipitate the behavior in which we are interested (1962:152).

Figure 4 gives the speed of the drum rhythms used in five Tsonga girls' initiation songs, as tape-recorded at the river-bank near Samarie, Northern Transvaal, in 1970, and musically transcribed in Johnston (1972:160, 178, 179, 185 and 190).

This figure shows that the accompanying drum-rhythms of these five Tsonga girls' initiation songs occur at the average rate of eight drum-beats (♪♩♪♩ ♩♪♩♪) every 1.0864 seconds, or just over 8 c.p.s. If this appears low compared to the range of human basic brain wave frequencies (8-13 c.p.s.), it should be pointed out that slightly lower

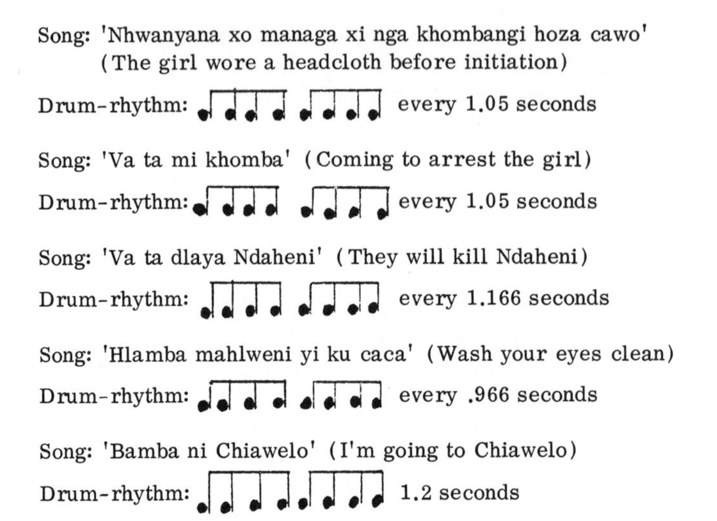

Figure 4. Proximity of Tsonga girls' initiation drum-rhythms to human basic brain wave frequency (8-13 c.p.s.)

frequencies favor sound stimulation, for the auditory area of the cortex exhibits the low-frequency theta rhythms.

Further to our discussion concerning changes in audition, it should be noted that, where two independent rhythm sources have been utilized in laboratory experiments, subject response was intensified: "For a few experiments, two light cources were used simultaneously with independent light frequencies ... the hallucinations described by subjects were of character so compelling that one subject was able to sketch them some weeks later" (Walter & Walter, 1949:63-64). In all of the Tsonga girls' initiation songs and dances that were tape-recorded, two or more independent drum, rattle, or clap patterns were used simultaneously — this is a well-known feature of most African music, and it may represent a psychophysiological 'net' evolved by adaptive cultural processes, functioning to ensnare individuals of different basic brain wave frequencies by the use of one integrative mechanism (Figure 5).

With regard to the remarkable sensory effect color-music synesthesia, we have noted that Tsonga initiation school novices are expected to see bluish-green color patterns ('mavalavala ya rihlaza') under the influence of the drug and during the playing of mime-specific drum-rhythms. The association of color and sound exists in many cultures. Among the Lau of the Solomon Islands "a low note is called 'bulu' (black), and a high note 'kwao' (white). These names are taken from charcoal marks made on a plank to indicate the tune: heavy down-strokes being 'black', and light up-strokes 'white'" (Ivens, n.d.:98; quoted by Merriam, 1964:96). In Mauritania, "within the different styles there are

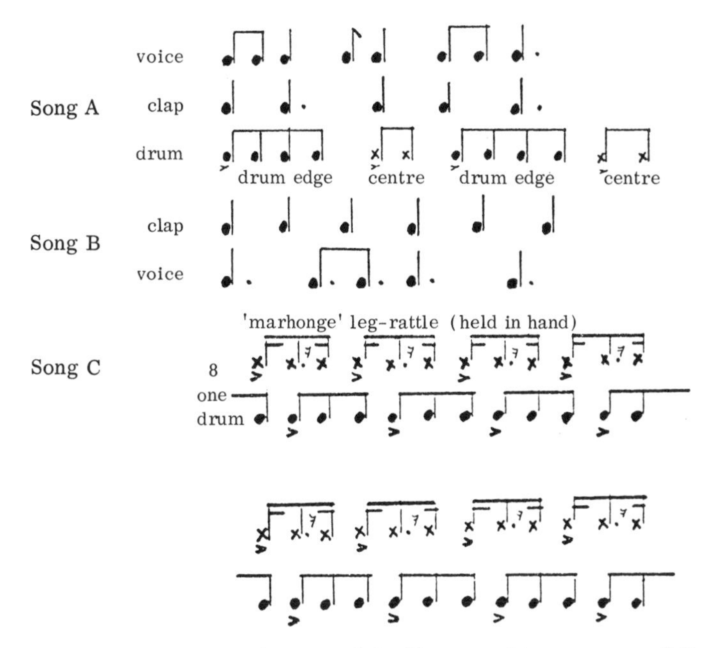

Figure 5. Examples A, B, and C. The use of two or more different rhythms simultaneously in Tsonga girls' initiation songs

two manners of performance — the 'white' and the 'black'. The latter, called 'Lekhal', is preferred by the Griot people since it allows more spectacular effects — the voice spanning a wide range and the notes being held longer ... the 'white' manner — or 'Lebiahd' — is favored by educated people. It is closer to the Arab tradition, and produces a smoother performance and more subtle singing" (Nikiprowetzky, 1961: n.p.; quoted by Merriam, 1964:97). In music in the West, certain instruments are considered to possess a 'dark' tone, usually low instruments such as the bassoon and cello. Trumpets are thought to have a 'bright' sound. Many keyboard artists think of the different possible tonalities in terms of colors. Scriabin wrote a major work during which a light-show was to be given.

In hallucinogenic drug experiments in the West, subjects often appear to perceive, emphasize, or select the 'hot' segment of the color are: reds, oranges. Among the Tsonga, drug-ingestion during initiation appears to result in perception of the 'cool' segment of the color arc: greens, blues.

The following list gives the color-perception responses of the twelve novices, as described by them at an interview some hours after the rite.

Color-perception responses of Tsonga initiation school novices under the influence of Datura and drumming

Novice	Her report	Translation
Novice A	tinyoka ya rihlaza	bluish-green snakes
Novice B	swivungu ya rihlaza	bluish-green worms
Novice C	tinyoka ya rihlaza	bluish-green snakes
Novice D	vilavila ya wasi	a blue whirlpool
Novice E	xihlovo ya wasi	a blue spring (water)
Novice F	mindzhati ya rihlaza	bluish-green lines
Novice G	tinyoka ya rihlaza	bluish-green snakes
Novice H	swirhendzewutani ya rihlaza	bluish-green balls
Novice I	mindhendzewuti ya rihlaza	bluish-green circles
Novice J	ndzhandzheni	the opposite river bank
Novice K	ndzhandza	the river bank
Novice L	tinyoka ya rihlaza	bluish-green snakes

Blue-green is the color of the harmless snakes found under the eaves of Tsonga huts, and which are thought to be ancestor-gods. Water in dreams is customarily associated with fertility. As for the lines, balls, and circles, hallucinogens enchance visual effects by changing retinal image and permitting the appearance of geometric forms and patterns. These visions are the physiological structures in one's own visual system, including the lattices, cones, cylinders, and other geometrics, suddenly amenable to observation.

Concerning distinction between blue and green, the Tsonga appear to be vague about it; it is interesting to note that the Kamayura Indians of Brazil use one word for the two, meaning parakeet-colored. Within the Tsonga environment green is certainly the more meaningful color, and the term 'rihlaza' is a catch-all for green and bluish-green objects.

Tsonga emphasis upon perceiving snakes may be compared to Zulu dream-interpretation: "The snake and 'tokoloshe' dreams are frankly sexual in character — and are often of violent sexual attack by either of these. The local interpretation of the snake dream is frequently: this means that there is a man that you fear" (Lee, 1969:322). The Tsonga possess many myths associating snakes with water. Water-fowl, for instance, are reputed to store small green snakes in their nests. The snake-like kudu antelope horn which is used to signal dance and mime directions, is repeatedly flushed with water during use. The selection of a riverbank site for performing the 'khomba' initiation rites must be done with care by an exegete, for fear of offending the snakes which are thereabouts.

The perception of the Tsonga bluish-green fertility snake and the hearing of the voice of the fertility god is largely stimulated by the suggestions of the officiant, who is knowledgeable in Tsonga ritualistic lore and in the psychological manipulation necessary to ensure group conformity during the rites. With the enforced diminution of the novices' critical faculties there is a decrease in reality testing, bringing about the compensatory need to seek support and guidance. The dissolution of self boundaries diminishes primary process thinking (doubting, etc.), external suggestions assume a concrete reality, and a supramotivational state ensues. The officiant becomes the silent inner voice. Manipulating the novices with powerful music, she leads them into the various consecutive mimes and dances with ther soundings upon the kudu horn. It is she who suggests the music-color association, the hearing of voices, and the fertility vision.

CONTRIBUTIVE EFFECT OF THE DANCING

In Tsonga dancing, the feet stamping at one rhythm, the hands clap at a second rhythm, and the voice sings syllables which float at a third rhythm over the accompaniment 'grid'. The hips, shoulders, and head meanwhile sway and undulate in time to various aspects of the music in a typically African series of gestures and postures. Neher states that "violent dancing and gestures make hyperventilation a possibility and increase the production of adrenaline, as well as cause a decrease in blood glucose which is used for energy" (1962:157). It has been shown that low blood glucose and the production of adrenaline increase susceptitbility to tapping of the basic brain wave rhythms (Strauss et al., 1952).

Complementary to the drug's effects are the auditory driving, the dancing, tactile reinforcement (beating with switches), visual effects (swaddling in see-through colored blankets), the suggestions of the officiant, and the cultural imperatives which associate dreams with the fertility-god and fertility with social success.

PRIMARY SCHOOL EDUCATION AND BELIEF IN WITCHCRAFT

All of the novices involved were senior students at Samarie Primary School, utilizing textbooks in English, Afrikaans and Tsonga, and studying modern scientific subjects such as (elementary) biology. They nevertheless believe firmly in the power of witchcraft, and greatly fear barrenness by witchcraft. One of the key dances in the 'khomba' rites involves a posture where the novices stand in file extending one arm and one leg, to receive consecrated amulets guaranteeing immunity from witchcraft curses. Such curses may emanate from within one's closest social circle, by reason of envy or just plain malice. A common source is one's co-wife, jealous of the husband's temporarily preferred sleeping companion. The Tsonga sing many witchcraft songs:

Call:	Hey, you witches
Response:	I really shall not live
Call:	My grandmother there at the old place
Response:	I really shall not live
Call:	Has left the village on that account
Response:	I really shall not live
Call:	Wash your eyes clean
Response:	I really shall not live
Call:	We told them, we shall not be spared

Here, "We shall not be spared" refers to the danger of barrenness by witchcraft, and "Wash your eyes clean" refers to both the fact that women must watch out for witchcraft in their husband's village, and to the fact that the drug-ingestion 'washes one's eyes', facilitating perception of the fertility-god.

The strong present-day belief in witchcraft, by adolescents attending a European-type school, suggests that such beliefs may not be dissonant with modernization and on-going social change. Jahoda's study of supernatural beliefs and changing cognitive structures among Ghanian university students found "coexistence between African and Western ideas and beliefs" (1970:115-130). In the parallel but different case of drug-ingestion, however, this state is difficult to envisage, for South African laws override native law in declaring the practice illegal.

AN ECOLOGICAL AND ENVIRONMENTAL EXPLANATION OF 'KHOMBA'

Much searching of the records of the Bantu Administration failed to turn up reliable figures of Tsonga infertility and infant mortality. We are dealing here with a rural, traditional society where malnutrition and diseases such as syphilis are sometimes prevalent and where it is not generally determined whether the husband or the wife is the infertile partner, where infants are not considered fully 'human' (recognized members of the society) until after the first year, for fear of their early death. Bilharzia and gastroenteritis claim some victims during this period. The closest approach to reliable figures came from the Swiss-manned clinic and mission at Elim Hospital, near Sibasa in the Northern Transvaal, where doctors' estimates put infertility in Tsonga women at about 30 per cent and infant mortality during the first year at about 35 per cent. Deceased infants are quietly buried, often without being reported. A well-known Tsonga song likens healthy infants to a true-ringing, well-baked clay pot that has survived the crucial firing process, and tells how 'weaklings' are buried in a broken clay pot, "for they cracked in the firing".

What is certain is that Tsonga bride who proves childless is in great trouble. It is not merely that the cattle which have been paid for her must be returned; often, the cattle have already been spent to bring the girl's brother a wife, and a chain reaction is involved. The barren woman may expect a lifetime of disgrace and working for others. To avoid the consequences of believed sterility, 'khomba' laws are strict. A novice declaring inability to perceive the required visions has, in the past, been known to receive supplementary Datura, thus ending heretic doubts.

DRUG-USE IN THE TSONGA EXORCISM RITES

Probably rooted in Tsonga history and migration, the Tsonga fear of Zulu ancestor-spirits and Ndau ancestor-spirits is manifested today in the phenomenon called 'mancomane', spirit-exorcism. Undesirable alien spirits possessing Tsonga individuals are commonly diagnosed as either of Zulu origin (the 'mandhlosi' spirits) or of Ndau origin (the 'xidzimba' and 'ziNdau' spirits); they are expelled via drug-use and the performance of spirit-specific music. Both the scale and the rhythm of the performed music is tailored to the nationality of the possessing spirit, as is the language of the song-words. The complex of interesting cognitive distinctions made by the Tsonga in their classification of spirit-associated phenomena reflects not only tribal history as it actually occurred, but Tsonga conceptions of what occurred.

The first signs of possession are a nervous crisis, chest pain, thinness, or excessive yawning. The set of sixty-four divining bones are thrown to determine the possibility of successful exorcism. If the four seashells among the bones should fall on their back, with the opening facing upward, the spirit will come out. If they fall face downward, the ceremony will be in vain.

The exorcist covers the patient with a blanket, burns *Datura fastuosa* roots, seeds, leaves, stalks, and flowers in a circle of embers under the blanket, and, together with his assistants (often his wives — the display of many wives constitutes a sign of previous successful practice), commences the prolonged loud, drumming on the tambourine-drums. An interesting aspect of this drum is that it is not indigenous to any other southern African people, but is found in the same shape, size, and function among Siberian, Guatemalan, Northern Irish, and other folk-healers. This may be due to its light weight, its simplicity of manufacture, and its solar appearance. The drumming is fast, around 9-12 c.p.s.

When the effects of the drug and the drumming (along with hypersuggestibility, tactile reinforcement, and heat-exposure) take their toll, the patient reveals the spirit's origin by singing. Possession by undesirable alien Zulu spirits ('mandhlosi') is revealed by singing songs containing the names of 19th-century Zulu conquerors of the Tsonga, such as Nghunghunyane:

Possessing spirit	Language of songwords	Melodic pattern	Rhythmic pattern
'mandlhosi'	Zulu	pentatonic	
'xidzimba'	Shona or Ndau	heptatonic	
'xiNdau'	Ndau or Rotse	heptatonic	

Figure 6. Spirit-specific characteristics of the music accompanying exorcism drug-use

Song 1

Call:	Nghunghunyane, personification of our people!
Response:	We name him the Ideal One!
Call:	We are his followers / His son neglected us / We never catch sight of him.

The diagonal strokes represent a repeat of the Response.

Song 2

Call:	Nghunghunyane was killed in the bush.
Response:	His enemies remain dancing.

To expel the Zulu spirit the exorcist will then order his drummers to play the appropriate 'mandhlosi' rhythm; should the spirit be revealed to be of Shona or of Ndau origin, there are special rhythms for each case (Figure 6).

In addition to spirit-specific rhythms, spirit-specific scale-patterns must permeate the melodies of the exorcism songs. The pentatonic (5-tone) scale is appropriate for Zulu spirits, and the heptatonic (7-tone) scale for Shona and Ndau spirits; this is confirmed by an interval-count of fifty-one 'mandhlosi' songs and twenty-four 'xidzimba' and 'xiNdau' songs (Figure 7). The heptatonic songs contain 8.5 per cent minor 2nds, which interval does not occur in Tsonga pentatonic music.

Following successful exorcism of the undesirable alien spirits, the patient is led to a spot under a special tree called the 'gandzelo', where his (or her) head is splashed with treated water from a large ceremonial seashell, and he is required to consume a small pellet of human fat or powdered bone. The throat is then tickled with a feather, inducing vomiturition, which is thought to appease the spirit.

Then follows a long period of rehabilitation and propitiation, indicating the transitional nature of the rite and the Tsonga fear of regressing (because of negligence) to one's former state. Tsonga individuals who have been through possession evince a much higher degree of religiosity than those who merely placate their own ancestor-spirits. For com-

Interval preference in 51 pentatonic exorcism songs

Interval (total of 1014 intervals)	% (approx.)
major 2nd, descending	29
minor 3rd, descending	21
major 2nd, ascending	15.5
minor 3rd, ascending	9
4th, descending	8.5
4th, ascending	5
major 3rd, descending	3
5th, descending	3
minor 7th, ascending	1.5
major 3rd, ascending	1.5
8ve, ascending	1
5th, ascending	1
major 6th, ascending	0.5
major 9th, ascending	0.5
	100%

Interval preference in 24 heptatonic exorcism songs

Interval (total of 496 intervals)	% (approx.)
major 2nd, descending	35
minor 3rd, descending	15
minor 2nd, descending	8.5
major 2nd, ascending	7.5
minor 3rd, ascending	6.5
major 3rd, descending	5.5
4th, descending	5
4th, ascending	4
minor 2nd, ascending	3.5
5th, ascending	2.5
minor 7th, ascending	2
major 3rd, ascending	2
5th, descending	1.5
major 6th, ascending	1
major 7th, ascending	0.5
	100%

Figure 7. Table showing cognitive distinctions made by the Tsonga between scales suitable for expelling Zulu spirits from drugged patients, and those suitable for expelling Shona and Ndau spirits (these last two require heptatonicism)

parative information on Tsonga exorcism in the early 20th century. consult Junod (1927, 2:479-504).

Tsonga exorcism, involving belief in spirits and other aspects of the supernatural, is a part of the overall Tsonga religious system, and forms part of an important dichotomy in which family spirits are at the other pole. Figure 8 shows the role of drug-use and of drumming in different contexts involving ancestrolatry and spirit exorcism.

CONCLUSION

Each of the ritual situations described is a rite of intensification, a tripartite sequence of separation — liminality — reincorporation of the group of novices, involving progressive stages of socially recognized (by the Tsonga) identity. The establishment of group social identity appears to take precedence over that of individual identity, which is one important difference between Tsonga society and Western society. Each of the biochemical, psychophysiological, musical, and cultural mechanisms utilized by the officiants at each stage of the rites of initiation and of exorcism, appear to be directed toward a group status-defining goal. Although some of the mechanisms employed have long

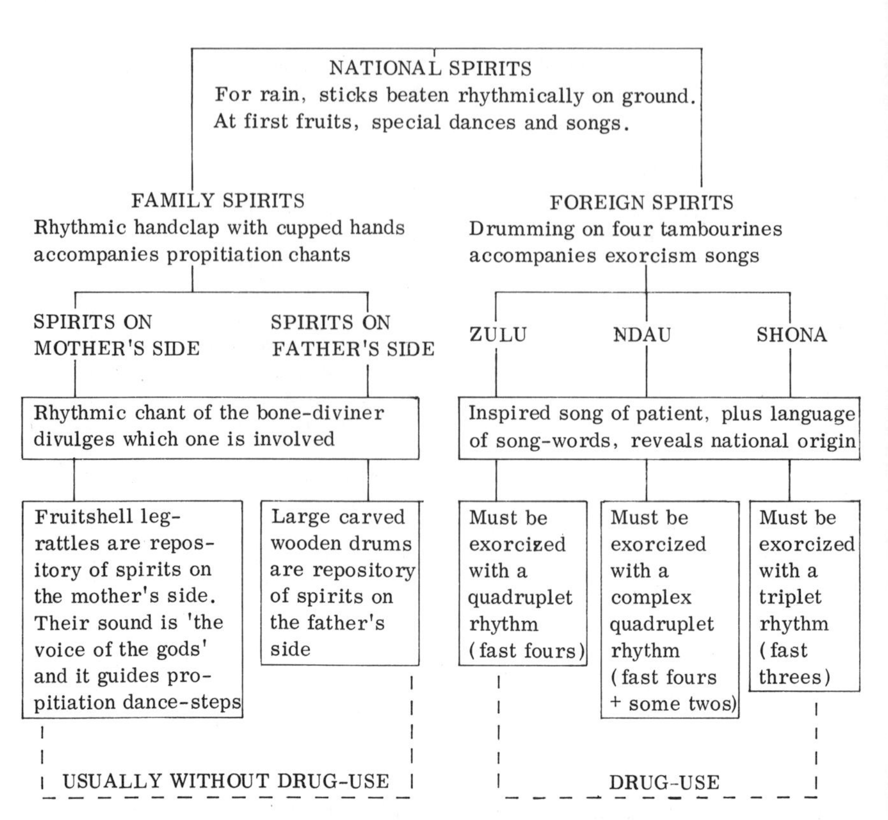

Figure 8. The role of music in the Tsonga religious system

been known rather superficially to explorers, missionaries, and early anthropologists as typical African phenomena (drumming, dancing), their origin and function have only recently come to be more fully understood. Too often in the past, African sociomusical behavior has been attributed to some mystical, hedonistic, or eudemonistic pleasure-search or lack of a sense of ethical responsibility. In fact, African drug-use, drumming, dancing, and singing are pragmatic and problem-solving. They are a culture's adaptive response to ecological and environmental pressures, and their particular mode of manifestation is determined by deep-rooted historical and psychological factors. Specifically, the availability of timber in the great wet rain forests of Africa, where drummers fell an entire tree for one drum, and the availability of tropical and subtropical plants containing hallucinogens, are complementary botanical assets of which the African has long taken advantage, and the use of which has subsequently diffused southward. The ecosystem which facilely yields drums and drugs, also generally yields tropical and subtropical ailments causing widespread sterility, hence the use of one to counter the effects of the other. These aboriginal cultural adaptations have throughout history been variously modified by contact and influence from encroaching alien cultures, so that there exists a complex of historical, social, and psychological forces inhibiting the continuance of traditional drug-use practices, and an opposing complex encouraging them, as follows.

A. Those tending to inhibit traditional drug-use practices

1. Labor migration and the acquisition of new values.
2. Government schools and European teachings.
3. Religious broadcasts from urban radio stations.
4. The use of imported religious phonograph records.
5. The disapproval of European-manned Christian churches.
6. The disapproval of (some) African Christian churches.
7. Attendance at government health clinics.
8. The close presence of the (prestigious) European farmer.
9. The introduction of new medicines such as aspirin.
10. White government legislation banning the free use of hallucinogens.

B. Those tending to encourage traditional drug-use practices

1. The existing linguistic code referring to magical potions.
2. Its strong identification with traditional social activities.
3. The great utility of many folk-healing practices.
4. Inculcation of children via folktale-telling.
5. Inculcation of traditional values via initiation schools.
6. The still strong authority of traditional herbalists.
7. The still strong kinship ties and hence tradition-oriented hierarchies.
8. The strong traditionalism of women, who are the culture-guardians.
9. Price in tribal origin.
10. Deliberate native preservationism, by African intellectuals.
11. Deliberate anthropological preservationism by European scholars.
12. The collection and dissemination of herbal practices by Asian traders operating herbal stores.
13. The dynamic renewal process by intertribal diffusion of medicinal and religious practices.

Not the least of the social problems which arise from drug-use by different cultural groups is the legal question. To relate this paper to the situation in the United States, it is sufficient to point out that, to the peyote-users of the Native American Church, government drug-control laws are oppressive, while, to the narcotics bureau, Indian drug-use practices represent a violation of national law. In Africa, the remoteness of many areas precludes strict law enforcement, but this condition may not prevail indefinitely, and African governments may soon have to face problems similar to those found in America.

BIBLIOGRAPHY

Balandrier, Georges, 1963, Sociologie actuelle de l'Afrique noire. Paris: Presses Universitaires de France.

Barber, T. X., 1970, LSD, marihuana, yoga, and hypnosis. Chicago:Aldine.

Blum, R., et al., 1964, Utopiates: the use and users of LSD-25. New York: Atherton Press.

Chapple, E. D. & G. S. Coon, 1942, Rites of intensification. In Principles of Anthropology. New York: Holt.

Fernandez, James W., 1965, Symbolic consensus in a fang reformative cult. American Anthropologist, 67:902-929.

Guttmann, E., 1936, Artificial psychoses produced by mescaline. Journal of Mental Science, 82:203-221.

Hartman, A. & L. Hollister, 1963, Effect of mescaline, LSD, and psilocybin on color perception. Psychopharmacology, 4:441-451.

Ivens, W. G., n.d., The island builders of the Pacific. Philadelphia: Lippincott.

Jahoda, Gustav, 1970, Supernatural beliefs and changing cognitive structures among Ghanian university students. Journal of Cross-Cultural Psychology, 1, 2(June):115-130.

Johnston, Thomas F., 1972, The music of the Shangana-Tsonga. Unpublished Ph.D. thesis for the University of the Witwatersrand, Johannesburg.

Junod, Henri A., 1927, The life of a South African tribe (2 vols.). London: Macmillan.

Klee, B., 1963, LSD-25 and ego functions. Arch.gen.Psychiatry, 8:461-474.

Leach, E. R., 1968, Ritual. In International Encyclopedia of the Social Sciences, Vol.13. New York: Macmillan.

Lee, S. G., 1969, Social influences in Zulu dreaming. In D. R. Price Williams (ed.), Cross-cultural Studies. Middlesex: Penguin Books.

Masters, R. E. L. & Jean Houston, 1966, The varieties of psychedelic experience. New York: Dell.

Merriam, Alan P., 1964, The anthropology of music. Evanston: Northwestern University Press.

Neher, Andrew, 1962, A physiological explanation of unusual behavior in ceremonies involving drums. Human Biology, 34(2):151-160.

Nikiprowetzky, Tolia, 1961, La musique de la Mauritanie. Paris: SORAFOM.

Pope, Harrison G., 1969, Tabernanthe Iboga: an African narcotic plant of social importance. Economic Botany, 23(2):174-184.

Slotkin, James, 1956, The peyote religion. Glencoe: Free Press.

Strauss, H., M. Ostow & L. Greenstein, 1952, Diagnostic EEG. New York: Grune & Stratton.

Swiderski, Stanislaw, 1965, Le Bwiti, société d'initiation chez les Apindji au Gabon. Anthropos, 60:541-551.

Tart, Charles T. (ed.), 1969, Altered states of consciousness. New York: Wiley.

Turner, Victor, 1967, The forest of symbols: aspects of Ndembu ritual. Ithaca: Cornell University Press.

Unger, S. M., 1963, Mescaline, LSD, psilocybin, and personality change. Psychiatry, 26:111-125.

Walter, V. J. & W. G. Walter, 1949, The central effects of rhythmic sensory stimulation. Electroencephalic Clinical Neurophysiology, 1:57-86.

MARLENE DOBKIN DE RIOS

13

PLANT HALLUCINOGENS, OUT-OF-BODY EXPERIENCES AND NEW WORLD MONUMENTAL EARTHWORKS

To the anthropologist interested in the problem of interpreting archaeological remains, particularly the belief systems and mythologies of now extinct peoples, New World massive earthworks present a fascinating problem worthy of consideration. Mounds containing burials and often enigmatic, massive symbolic earth formations difficult to interpret can be found in three major culture areas of the New World — the Adena/Hopewell peoples, the Olmec of the Gulf Coast of Mexico, and the Nazca peoples of southern Peru.

In my area of specialization, hallucinogens and culture, I have had occasion to conduct cross-cultural research on the use of such psychotropic plants, often prior to European contact in traditional societies of the world. In preparing a report for the U.S. National Commission on Marihuana and Drug Abuse (Dobkin de Rios, 1973a), I found common themes emerging from the ethnographic data among a number of world societies where such psychoactive plants have been used for magico-religious purposes. I would like to draw upon some of the generalizations available from this material in an attempt to make sense out of the content, structure and meaning of the New World massive earthworks.

In this paper, I will argue that due to shamanistic, out-of-body experiences, the so-called aerial voyage, prehistoric New World massive earthworks were constructed. The purpose of such expenditures of time and labor was to make known certain cosmological messages not only to supernatural forces, but to members of the community, as well as other shamans in conflict with the social group. To argue this point, I would like to evaluate the archaeological record, synthesize clinical studies available concerning hallucinogenic ingestion in patient and experimental populations and ensuant out-of-body experiences, analyze the cogent characteristics of shamanism generally pertinent to my thesis, and examine common, recurring themes of relevance linked to the use of psychotropic substances in traditional world societies.

In arguing these points, I must reiterate (as I have done so previously in print, cf. Dobkin de Rios 1972c, 1974a, 1975) the importance for the anthropological study of drug-induced states of consciousness in understanding cultural patterning of visionary experience. As my researches in this area have shown, when an individual belongs to a group where plant hallucinogens are commonly used in ritual performance, socialization brings forth cultural influences on drug-induced visions. I dwell on this issue here because the cultural patterning of consciousness will become important in my discussions of the possible shared meanings attachable to New World massive earthworks.

To begin the discussion, I would like to briefly summarize the magnitude and particulars of the North, Central and South American earthworks, and present documentation for the existence and availability, for use, of a series of hallucinogenic plants whose effects on man's central nervous system are patterned and discernible.[1]

NEW WORLD MASSIVE EARTHWORKS

1. Adena/Hopewell mound builders of North America

Throughout the mid-western United States in two closely-linked archaeological periods — Adena and Hopewell — we find hundreds of mounds and earthworks in a variety of forms. They are located as far afield as the Ohio River drainage valley, Kentucky, Indiana, West Virginia, Pennsylvania, Southern Ohio and Wisconsin. There are more than 5,000 mounds/earthworks among the Hopewell horizons alone, despite the fact that they are not generally associated with settlements (Shetrone, 1930:33). Needless to say, there are variations in form, types, burial patterns, etc. Common to this group, and reported in the extensive literature, is a discussion of early New World peoples, for whom agriculture is not highly developed. The Adena peoples, the earliest grouping, were hunters and fishermen and may have practiced some simple form of agriculture. The attribution of agriculture to this group, in fact, is predicated upon their construction of earthworks; archaeologists argue that such massive constructions could not have been erected without large work parties, supported by food surpluses which can be mustered only in agricultural societies. Among the Adena peoples, conical earthworks ranged from two to seventy feet high. Sacred enclosures were sometimes associated with the earthworks, which were circular in form and probably not fortified. Webb & Snow (1945, cited in Caldwell, 1964:6) suggest these may have functioned as meeting places for social subdivisions of a tribe. The sacred circles occurred in groups of two to eight. Mound burial practices, often associated with circular earthworks, were confined to a special class of prestigeful individuals, probably shamans, due to the types of funerary goods involved. For example, in the Ayers Mounds, discussed by Webb & Baby (1957:73), a shamanistic burial is suggested by the discovery of leather encasing portions of a skeleton with a male buried in what was probably the regalia of an animal skin; some use of medicine bags characterize Adena burials as well (ibid.:75). Animal masks have also been found in Adena burials which may have been widespread ceremonial symbols used by shamans (Dragoo, 1964:11). Among the Hopewell mounds, which have been viewed as developing from Adena culture, we find a large variety of effigies in the form of birds, animals, serpents, and geometric forms. The latter group has been suggested to represent cosmic symbols of a probable religious character. I will return to this topic later on.

Hopewell's geographic range of colossal geometric earthworks was enormous. It was characterized by the production of ceremonial objects intended for deposition with the dead (Silverberg, 1968:266). Although in the literature, Middle American influence has been suggested as having affected the Adena/Hopewell peoples (see especially Spaulding, 1952; Silverberg, 1968), there does not seem to be any concrete archeological evidence. Such discussions of Mesoamerican influence generally have focused on mounds such as the Turner Group in Ohio, where large amounts of copper artifacts include a horned serpent (Willoughby, 1922:95).

In discussing the Wisconsin Hopewell mounds, Shetrone has described the Lower Dells Group in Sauk County containing three birds flying in unison. The lowest bird has a wing spread of 240 feet. Shetrone points out that the custom of mound building is found almost entirely across the continent of Asia in extensive steppes and plains, which, of course, is the home of Paleo-Indian shamanistic religion. What Shetrone calls the effigy or image mounds are the ones to which I address myself in this paper. They were constructed in the form of birds, animals, serpents and human beings. They occurred in southern Wisconsin and adjacent areas, with two or three noticeable examples in Ohio. Shetrone reiterates that the effigy mounds were believed to have had their origin in sacred or religious observances and appear totemic in character, sometimes containing human burials. He does not connect the large mounds to the geometric enclosures of Ohio, although I would argue that they probably are part of the same complex (1930:29, 32-34). The latter groupings represent squares, circles, octagons, and crescents. In southern Ohio alone, geometric earthworks enclose up to 100 or more acres of land. The image forms occur in the plastic arts of the same cultural horizons, as well as on bone sculptures and ceramics.

Shamanistic elements, in my opinion, are widespread in these mounds of the Hopewell area. Much of the art forms show scenes which appear to be shamanistic in character. One must, however, be careful not to assume a facile explanation of the shaman battling supernatural forces. Rather, the militant nature of shamanism, discussed by Castaneda (1972), Furst (1965) and Sharon (1972a, 1972b) must be taken into consideration. From my own empirical studies of drug using societies in the Peruvian Amazon and Coast, (see Dobkin de Rios, 1968a, 1968b,

1970, 1972a, 1972b, 1972c, 1973b) I found much to my surprise that drug-using shamans spend as much time fighting other witches and evil-doers as they do in controlling spirit forces to do their bidding. One non-Hopewellian mound group in Georgia, the Etowah group described by Shetro (1930) depicts rampant warriors equipped with the paraphermalia of war, wearing antler headdresses, feather robes and other ceremonial accoutrements. One warrior has a severed head of a human victim in one hand, and a ceremonial axe or sword in the other. This may represent certain shamanistic themes argued for by Furst in west Mexican tomb art (1965) and by myself for Mochica art (n.d.) in northern Peru. Even Shetrone (1930: 28) has inquired if a shamanistic battle linked to supernatural forces was in question.

2. Olmec culture

In the rain forest of Mexico's Gulf Coast, an ancient people called the Olmec, dating from the first millennium B.C. are acknolwedged today to be the forerunners of the Maya and Aztec civilizations. They were the first American Indians to achieve as Coe writes (1971: 68) "the level of social, cultural and artistic complexity which permits us to call them civilized". They were probably organized into a coercive state by 1200 B.C., with control over large populations and over a fairly large area. They had a state religion which centered on a jaguar-like rain god (1968: 63). The Olmec developed an art style based on combined features of a snarling jaguar and human infant. Common, too, to their art is a jaguar monster mask and humans with jaguar face profiles. In one Gulf Coast archaeological site, San Lorenzo, Coe found a monumental earthwork three quarters of a mile long, reaching out like fingers on its north, west and south sides with long narrow ridges divided by ravines (1968: 45). A pair of ridges on the western side exhibit bilateral symmetry, with every feature on one ridge matched in mirror fashion by its counterpart. The ridges are artificial and consist of fill and cultural debris as deep as 25 feet. Coe suggest that this construction is a gigantic bird flying eastward, its extended wing feathers forming ridges on the north and south, with its tail trailing to the west (1971: 69). It is possible that the structure is aligned with some astronomical bearing. Coe comments that such a grandiose plan can only be appreciated from the air and suggests that an ancient Olmec ruler/priest, inspired by cosmological ideas, ordered this construction to impress the gods and men (ibid.: 70).

3. Nazca culture of southern Peru

In the southern desert coast of Peru, near the hills of the Rio Grande River, there is an important early Peruvian archaeological people, the Nazca, dated from about 1000 A.D. (Kubler, 1962: 284). It is known for some of the finest textiles and ceramics of ancient America. High in the elevated desert plains, a vast network consisting of thousands of straight lines and large dirt drawings on the ground has been recorded by scholars (Reiche, 1949; Kosok, 1947; Kosok & Reiche, 1949). Several hundred such earthworks exist, made by removing small stones that covered the desert from the area of the path. These stones were placed along the sides, and form slightly elevated ridges (Reiche, 1949: 207-208). Since no rain falls in this area of the world, preservation has been fairly continuous. The paths run always continuously, never crossing and complete the figure close to its beginning. Forms that are discernible include flying birds, insects, felines, killer whales, fish, spiral forms (possibly coiled snakes), snake-birds, flying pelicans and a large plant form. Some are over 1,700 meters long, while others extend over two and a half kilometers, Kosok, an archaeologist has suggested that these figures may have been made for walking on in festive occasions by priests and are probably sacred objects (1949: 212). All the earthworks motifs occur on the tapestries and ceramics of the region. Reiche, an astronomer and mathematician, suggests a variety of astronomical significances of these lines, which she documents as solstitial and equinoctial points upon the horizon (1949: 45).

Dating from the early Nazca period about 1000 A.D., the figures are always found closely associated with a large enclosure or wide road, whose explanation is given as that of totem-like symbols belonging to various kinship groups. Kubler writes that the function of the large curved figures of plants and animals of the effigies is unknown, but that their style is close to early Nazca drawings. He suggests images symbolic of constellations. To quote:

> It is perhaps unexpected, but it is not improper to call these lines, bands and effigies a kind of architecture. They are clearly monumental, serving as an immovable reminder that here an important activity once occurred. They inscribe a human meaning upon the hostile wastes of nature in a graphic record of a forgotten but once important ritual. They are an architecture of two-dimensional space, consecrated to human actions rather than to shelter and recording a correspondence between the earth and the universe ... They are an architecture of diagram and relation, with the substance reduced to a minimum (1962: 286).

It is interesting to note that the main analyses of these earthworks were done only after 1944, when aerial photography could be employed to map them. Kosok compares these huge drawings to the animal mounds of the northern U.S., just as Coe sees resemblances of his Olmec data to the Adena/Hopewell and Nazca mounds (1947: 295).

Now that I have sketched the particulars of the massive earthworks for these three cultures, I would like to turn to the available evidence for the presence of plant hallucinogens.

NEW WORLD PLANT HALLUCINOGENIC USE IN THE AREAS OF MASSIVE EARTHWORKS

As has been discussed elsewhere, New World Indian groups used a large variety of plant hallucinogens. Further, they most probably brought with them from their Asian homeland a propensity to experiment with such substances, especially in light of their early shamanistic religious values based on the primacy of personal revelation (La Barre, 1970). The reader is referred to Emboden's book, 'Narcotic plants' (1972) for encyclopedic materials on the data to follow.

In 1973, Janiger and I published an article entitled 'Suggestive hallucinogenic properties of tobacco'. In it, we argued that in the smoke condensates of the common *Nicotiana* species, two alkaloids called harman and norharman are formed, which are mildly hallucinogenic. In that article, we assembled data on Amerindian use of tobacco, which we suggested was used by native peoples as if it had psychotropic properties (see also Wilbert, 1972). Certainly, the Adena/Hopewell peoples heavily relied on the tobacco plant, attested to by the large number of pipes found in excavations. Shetrone (1930:126) cites one Ohio Hopewell mound where over 200 tobacco pipes were found, with representations of hawks, eagles, dogs, racoons and quails. As I have discussed elsewhere (Dobkin de Rios, n.d.), a common hallucinogenic-linked art motif concerns not only the representation of drug plants in the plastic arts, but themes connected to the magico-religious use of such plants. Animals such as those mentioned above could easily correspond to shamanistic animal familiars available to magical manipulation by drug-ingesting practitioners. As Pitt-Rivers has pointed out in a Central American study of applicability to much of New World drug art (1970), many New World peoples relate to animal species qua species in terms of particular attributes or qualities which explain certain cultures' hierarchical ranking of them in the world of nature. Thus, as Sharon has pointed out, the common use of the humingbird in Moche art (an area of the New World where on-going hallucinogenic rituals are commonplace) may be related to the characteristic of that bird to suck at the attar of various flowering plants. In a metaphoric sense, shamanistic healing in a wide number of New World societies entails the use of sucking the afflicted regions of the body to symbolically extract an intruded object, placed there by a rival sorcerer.

Returning again to the Hopewell data, Moorehead, in 1922, described an important copper form found in altars of one site, Mound City. Small mushrooms were discovered in connection with eagle and bird effigies (p.174). Some larger ones with longer stems were also present at this site, which he suggests portrayed the *Amanita* species, one variety of which in recent years Wasson has argued is strongly hallucinogenic (1957, 1968).[2] Once again, shamanistic animal familiar linked to drug ingestion may be in question. In addition, *Ilex cassine,* an infusion that causes a visionary state, is common throughout the southeast United States and Ohio Valley and is available for use (William Emboden, personal communication). Peyote, diffused northward from Mexico, is yet another candidate. A member of the nightshade family, *Datura stramonium,* is a highly potent plant which is used in this area in prehistoric times. Along the southern reaches of this culture during historical times, another hallucinogenic plant, the mescal bean and its cultic use has been recorded, dating back a couple of thousand years (*Sophora secundiflora;* cf. Howard, 1957).

Empirical evidence has accumulated to indicate that the use of such psychotropic plant substances is quite old in human history as opposed to the idea that their use indicates denigration of archaic shamanistic techniques as suggested by Eliade (1964). Such plant ingestion is linked to hunting and gathering peoples whose mythologies speak of animals revealing to man the secrets of the hallucinogenic plants (see Dobkin de Rios, 1973a). Siegel (1973) independently has confirmed this in his studies of animals' purposeful search and use of hallucinogenic plants.

If we turn to Mesoamerica and examine the Olmec data, we find an area where a large variety of plant hallucinogens exist and were available for use. To the north, among the Aztecs and of probable early origin, at least four major hallucinogenic plants were utilized, including the morning glory seed, *Ololiqui*, the peyote cactus (imported from the northern areas), *Datura* sps., and a large variety of psychotropic mushrooms. In addition, lesser known hallucinogenics, including members of the *Salvia* mint family were incorporated into magical rituals. In a recent paper (Dobkin de Rios, 1974b), I have argued that the Mayas, located to the south of the Olmec, utilized three psychotropic flora/fauna, including once again, the mushrooms, the waterlily plant (which contains apomorphine-like alkaloids)[2] and the toad. The latter creature contains bufotenine in its poison glands. Moreover, a tryptamine substance is contained in the skin of the common *Bufo alvarius* toad. Certainly, most Mesoamerican specialists lacking clear evidence in general about the parameters of Olmec civilization, nonetheless would agree that hallucinogenic plants played an important part in their social life (cf. Furst, 1968:162; 1972).

Among the third peoples of the New World under consideration, the Nazca, we find an area of Peru connected to the north coast region by native roads, which eventually become the Pan American highway in the coastal areas in recent decades. To the north, some 600 miles, a wide variety of hallucinogenic plants, which include some hallucinogenic cactii, were used in prehistoric times among the Moche, a people contemporary with the Nazca. The coca plant was known and widely chewed in Peru, both in the foothills and highland region. It was known in the coastal areas, as attested to in prehistoric pottery representations. It has been suggested that in high dosage levels, trance-like states are obtained. Palomillo has observed excessive dosage use, according to local highland Peruvian standards among contemporary Quechua-speaking shamans as part of their healing ritualistic activity (personal communication, Aquiles Palomillo). In the tropical rain forest, there are numerous hallucinogenic plants, including the *Banisteriopsis* sps., ayahuasca (which I have reported on), as well as *Datura* sps. Hallucinogenic snuff, documented archaeologically by Wassén to the south, and ethnographically to the east of the Nazca, additionally were available for use, as well as a potent form of tobacco, used even today in magical shamanistic healing rituals along the coastal region (see Dobkin de Rios, 1968a, 1968b, 1973b). In the three areas where massive earthworks are known, plant hallucinogens were readily available. From all the pertinent evidence, they were probably incorporated into the cultures in question.

THE OUT-OF-BODY EXPERIENCE

Having examined the evidence for plant hallucinogenic use co-occurring in regions of the New World where massive archaeological earthworks were constructed, I would like now to turn my attention to a commonly reported subjective effect of plant hallucinogens, the out-of-body experience. Green (1968) has labeled a particular effect achieved both spontaneously and through means of drugs as the ecosomatic effect, the commonly called out-of-body or aerial voyage experience. Throughout much of the non-scientific literature, this subjective state has been reported frequently as occurring spontaneously without any chemical intervention. Nonetheless, such experiences are a common feature of the hallucinogenic experience (cf. Hofmann, 1959; Masters & Houston, 1966; Tart, 1971), and are often referred to as the "depersonalization effect" (see Ludwig, 1969:14). Ludwig describes a wide array of distortions in body image in all altered states of consciousness as a schism between body and mind, or a dissolution of boundaries between self and other, the world, or the universe. Green defines the ecosomatic experience as "objects of perception organized in such a way that the observer seems to himself to be observing them from a point of view not coincident with his physical body" (1968:111). Another state which she titles the "asomatic state" is an ecosomatic state in which the subject is temporarily unaware of being associated with any body or spatial entity. Barber, in summarizing the clinical studies of LSD and other psychedelic drugs up to 1970, wrote of changes in body image as well. Practically all subjects state that their body ... "feels strange or funny. More bizarre feelings are registered at higher drug doses, such as a body melts into the background or floats in space" (1970:22).

A colleague of mine at the California College of Medicine at the University of California, Irvine, Dr. Roland Atkinson is currently involved in a research project examining the effects of nitrous oxide on man. He has summarized the literature on out-of-body experiences[3] and used the term autoscopy, which is a general term meaning the perception, usually visual, of one's own physical body in objective space. He uses the term external autoscopy for the perception of one's own body in objective space from the out-of-body experience. Atkinson argues that a subject may respond to heightened awareness as pleasant or instructive, particularly if he seeks it out through training and repeated efforts, including hallucinogenic drugs. Ranging on a continuum of effects, the out-of-body experience includes effects where the subject sees his physical body in objective space but experiences no body-like container encompassing the external locus of the subject's awareness. The term 'self' rather than body is more appropriate in this context. Reports at this level usually emphasize the other-wordly, mystical and paranormal nature of perception and experience. This type of experience tends to be associated with profound drug-induced altered states of consciousness.

Atkinson is interested in the neurological and psychiatric dimensions involved in the production of such states, a topic outside the scope of this paper. However, this discussion should illustrate the relationship of drug-induction and the subjective effects of the out-of-body experience, which I would like to link to the shamanistic ingestion of plant hallucinogens among the Adena/Hopewell, Olmec and Nazca peoples.

As Eliade documents, ecosomatic experiences, either drug-induced or not, have been reported with shamanistic religion. He has summarized the aerial voyage as follows:

> Siberian, Eskimo and North American shamans fly. All over the world, the same magical power is credited to sorcerers and medicine men ... All this makes us think of the ornithomorphic symbolism of the Siberian shamans' costumes ... According to many traditions, the power of flight extended to all men in the mythical age; all could reach heaven whether on wings of a fabulous bird or on the clouds ... We should make it clear, however, that here such powers often take on a purely spiritual character; flight expresses only intelligence understanding of secret things or metaphysical truths ... Magical flight is the expression of both the soul's autonomy and of ecstasy. The myth of the soul contains in embryo a whole metaphysic of man's spiritual autonomy and freedom ... The point of primary importance here is that the mythology and the rites of magical flight peculiar to shamans and sorcerers confirm and proclaim their transcendence in respect to the human condition; by flying into the air, in bird form or in their normal human shape, shamans proclaim the degeneration of humanity (1958:480-481).

DISCUSSION

I believe that the foregoing evidence indicates that in these three areas of the primitive world where plant hallucinogens were probably used, drug-induced shamanistic out-of-body experiences were provoked for magico-religious ends. To link this phenomenon, however, with the image mounds and earthworks of the Adena/Hopewell, Olmec and Nazca peoples requires a discussion of features of shamanism relevant to the data assembled earlier. I would particularly like to examine the role of the shaman as the psychopomp, the spiritual guardian of a community who is obliged to confront and combat his group's adversaries. A major part of his activity includes healing disease and neutralizing those misfortunes which have occurred to members of the community through the machinations of enemies. Shamans are famous for their ability to transform themselves into powerful animal figures (mentioned earlier), the familiars or 'naguals' in Pitt-Rivers' terminology, whom they send to do their bidding, to rectify evil or redress the harm caused their clients. These familiars are generally chosen because of some characteristic which, through sympathetic magic, are believed controllable by the shaman. In my opinion, the effigies of animals found throughout the New World massive earthworks represent these shamanic familiars.

Eliade, once again, has summarized the vast literature on shamanism with regard to this point as follows:

> It is the shaman who turns himself into an animal ... The shaman becomes an animal -spirit and speaks, sings or flies like the animals and birds. We must take into account the mystical solidarity between man and animal which is a dominant characteristic of the religion of the paleohunters (1958:93).

I would argue that the drug-using shaman, steeped in a religion of hunters, with spirit familiars on call to serve him, has a subjective experience which includes the sensation of flying. My point is quite simple. One need not fly in the air to really fly. Thus, the New World massive earthworks, difficult for the Westerner to conceptualize visually outside of an airplane voyage, are perhaps more simply explicable as the projection by the shaman of the animal or totem familiar from the heights of ecstasy through which he soars.

Geometrics in the earthworks, on the other hand, may be linked to another common LSD-like effect. Here, I am speaking of the frequently reported geometric forms, the kaleidoscopic visionary patterns reported by drug users, which Barber (1970) has argued may represent a change in retinal structure. Thus, a drug-user is seeing through the geometric forms of his eye, a near universal experience for those using hallucinogens. Others like Arguellas (1972:20) have mentioned reports of the universality of the Mandala form, a square encircled, in the art of both the Old and New World peoples, somewhat less easy to substantiate.

Given the arguments for the militant nature of the shaman, protecting his community against the evildoings of others, as well as his role as intermediary with the supernatural, such monumental earthworks in this analysis are viewed as constructed to warn rival shamans of the powers that were controlled by the psychopomp of a given area, to re-affirm supernatural contact and maintain social solidarity. Enormous expenditures of labor and cooperation, needed to construct such earthworks (extending perhaps from generation to generation), re-affirmed the bonds that link men together. Ties of cooperation maintain intra-group harmony and are important in small-scale societies. Finally, in the cases discussed in this article, the symbolic forms of the image mounds consisted of elements of ritualized belief already present in the plastic arts of the particular cultures in question.

As an emblem of power, constructed in the symbolic idiom of each culture, the monumental earthworks are truly worthy of our interest in the area of shamanistic religion and hallucinogenic use.

NOTES

1. See Ludwig (1969:13-16) for a detailed discussion of psychological and somatic effects of altered states of consciousness, including those induced by hallucinogens.
2. It is important to note here that no actual clinical evidence has been recorded on the psychotropic effects of the *Amanita* sps. at the time of this writing.
3. Out-of-the-body experiences: a critical review of the phenomenon and discussion of theoretically possible mechanisms for its occurrence (unpublished manuscript).

BIBLIOGRAPHY

Arguelles, José & Miriam Arguelles, 1972, Mandala. Berkeley: Shambhala.

Barber, Theodore X., 1970, LSD, marihuana, yoga and hypnosis. Chicago: Aldine.

Castaneda, Carlos, 1972, Journey to Ixtlan. New York: Simon & Schuster.

Coe, Michael D., 1968, San Lorenzo and the Olmec civilization. In Elizabeth P. Benson (ed.), Dumbarton Oaks Research Library and Collection for Harvard University. Washington, D.C.

Coe, Michael D., 1971, The shadow of the Olmecs. Horizon, 13:67-74.

Dobkin de Rios, Marlene, 1968a, *Trichocereus-pachanoi* — a mescaline cactus used in folk healing in Peru. Economic Botany, 22:191-194.

Dobkin de Rios, Marlene, 1968b, Folk curing with a psychedelic cactus in northern Peru. International Journal of Social Psychiatry, 15:23-32.

Dobkin de Rios, Marlene, 1970, *Banisteriopsis* used in witchcraft and folk healing in Iquitos, Peru. Economic Botany, 24:296-300.

Dobkin de Rios, Marlene, 1972a, Visionary vine: psychedelic healing in the Peruvian Amazon. San Francisco: Chandler.

Dobkin de Rios, Marlene, 1972b, Curing with Ayahuasca in a Peruvian Amazon slum. In Michael J. Harner (ed.), Hallucinogens and shamanism. New York: Oxford.
Dobkin de Rios, Marlene, 1973a, The non-western use of hallucinogenic agents. In Drug use in America: problem in perspective, Vol. I, Appendix. 2nd Report of the National Commission on Marihuana and Drug Use, I:1179-1235.
Dobkin de Rios, Marlene, 1973b, Peruvian hallucinogenic folk healing: an overview. In Ramon de la Fuente & Maxwell Weisman (eds.), Psychiatry. Proceedings of the 5th World Congress of Psychiatry. Amsterdam: Excerpta Medica.
Dobkin de Rios, Marlene, 1974a, Cultural persona in drug-induced altered states of consciousness. In Thomas Fitzgerald (ed.), Social and cultural identity: problems of persistence and change. Southern Anthropological Society Proceedings No. 8. Athens, Georgia: University of Georgia Press.
Dobkin de Rios, Marlene, 1974b, The influence of psychotropic flora and fauna on Maya religion. Current Anthropology, 15:147-164.
Dobkin de Rios, Marlene, 1975, Man, culture and hallucinogens: an overview. In Vera Rubin (ed.), Cannabis and culture. The Hague: Mouton.
Dobkin de Rios, Marlene, n.d., Plant hallucinogens and the religion of the Moche — an ancient Peruvian people (unpublished manuscript).
Dragoo, Don W., 1964, The development of Adena culture and its role in the formation of the Ohio Hopewell. In Joseph Caldwell & Robert L. Hall (eds.), Hopewellian studies. Illinois State Museum Scientific Papers 12.
Eliade, Mircea, 1958, Shamanism: archaic techniques of ecstasy. Trans. W. Trask. New York: Pantheon.
Emboden, William, 1972, Narcotic plants. New York: Macmillan.
Furst, Peter, 1965, West Mexican tomb sculpture as evidence for shamanism in prehistoric Mesoamerica. Anthropologia, 15:29-60.
Furst, Peter, 1968, The Olmec were-jaguar motif in the light of ethnographic reality. In Elizabeth P. Benson (ed.), Dumbarton Oaks Research Library and Collection for Harvard University Conference on the Olmec. Washington, D.C.
Furst, Peter, 1972, Ritual use of hallucinogens in Mesoamerica: new evidence for snuffing from the preclassic and early classic. In Religion en Mesoamerica XII, Mesa Redonda, Sociedad Mexicana de Antropologia.
Green, Celia, 1968, Out-of-the-body experience. New York: Ballantine Books.
Hofmann, A., 1959, Psychotomimetic drugs, chemical and pharmacological aspects. Acta Physiol. Pharmacol. Neerl., 8:240-258.
Howard, J. H., 1957, The mescal bean cult of the Central and Southern Plains: an ancestor of the peyote cult? American Anthropologist, 59: 75-87.
Janiger, Oscar & Marlene Dobkin de Rios, 1973, Suggestive hallucinogenic properties of tobacco. Medical Anthropology Newsletter, 4:6-11.
Kosok, Paul, 1947, The mysterious markings of Nazca. Natural History, 56: 200-209.
Kosok, Paul & Marie Reiche, 1949, Ancient drawings in the desert of Peru. Archaeology, 11:206-215.
Kubler, George, 1962, The art and architecture of ancient America. New York: Penguin.
La Barre, Weston, 1970, Old and New World Narcotics: a statistical question and an ethnological reply. Economic Botany, 24:73-80.

Ludwig, Arnold, 1969, Altered states of consciousness. In Charles Tart (ed.), Altered states of consciousness. New York: Wiley.
Masters, R. E. L. & Jean Houston, 1966, The varieties of psychedelic experience. New York: Dell.
Moorehead, Warren K., 1922, The Hopewell Mound Group of Ohio. Museum of Natural History Publication 211:6.
Pitt-Rivers, Julian, 1970, Spiritual power in Central America. The Naguals of Chiapas. In Mary Douglas (ed.), Witchcraft confessions and accusations. New York: Tavistock Publications.
Reiche, Maria, 1949, Los dibujos gigantescos en el suelo de las Pampas de Nasca y Palpa. Lima.
Sharon, Douglas, 1972a, Eduardo the Healer. Natural History, 81:32-47.
Sharon, Douglas, 1972b, The San Pedro cactus in Peruvian folk healing. In Peter Furst (ed.), Flesh of the Gods: ritual use of hallucinogens. New York: Praeger.
Shetrone, Henry Clyde, 1930, The Mound builders. New York: Kennikat.
Siegel, Ronald K., 1973, An ethnological search for self-administration of hallucinogens. The International Journal of the Addictions, 8:373-393.
Silverberg, Robert, 1968, Mound builders of ancient America. Greenwich, Conn: NY Graphic Society.
Spaulding, Albert C., 1952, The origin of the Adena culture of the Ohio Valley. Southwestern Journal of Anthropology, 8:260-268.
Tart, Charles T., 1971, On being stoned: a psychological study of marihuana intoxication. Palo Alto, California: Science and Behavior Books.
Wassén, S. Henry, 1965, The use of some specific kinds of South American Indian snuff and related paraphernalia. Etnologiska Studier, 28.
Wassén, S. Henry, 1967, Anthropological survey of the use of South American snuffs. In Daniel H. Efron (ed.), Ethnopharmacologic search for psychoactive drugs. Washington, D.C.
Wasson, R. Gordon & V. P. Wasson, 1957, Russia, mushrooms and history, 2 volumes. New York: Pantheon.
Wasson, R. Gordon, 1968, Soma, divine mushroom of immortality. New York: Harcourt, Brace & World.
Webb, William S. & Raymond S. Baby, 1957, The Adena People No. 2. Ohio Historical Society, Ohio State University Press.
Wilbert, Johannes, 1972, Tobacco and shamanistic ecstasy among the Warao Indians of Venezuela. In Peter Furst (ed.), Flesh of the Gods: The ritual use of hallucinogens. New York: Praeger.
Willoughby, Charles C., 1922, The Turner Group of Earthworks. Hamilton County, Ohio. Peabody Museum of American Archaeology and Ethnology. Cambridge, Mass.

FUTURE RESEARCH

Having looked at a variety of studies dealing with drug use, many of which were ethnobotanically based and hallucinatory in nature, it would be of value to look to the future. Under this rubric we refer to methods of research and subjects for research.

In the first chapter in this section Sorenson reports on the development of visual anthropology within the National Anthropological Film Center. He argues that such undifferentiated raw data are "richer and less skewed than culture-bound verbal description of the same event". While we might not all be in a position to utilize this methodology, particularly in the field of so delicate a subject as drug use, the method does have great value.

The last chapter returns to the richness of human discovery and inventiveness, the area of plant hallucinogens. Schultes concentrates this paper on the New World, partly due to the greater numbers of plants used in this part of the world. Elsewhere this author pointed out that between 90-100 species of hallucinogenic plants are employed in the Western Hemisphere as against only about a dozen in the Eastern (Schultes, 1973). The fact that most of those in the West occur in the New World is confirmed in this chapter.

While we commended anthropologists for getting into the center of modern urban and government studies, we strongly agree with Schultes' call that we pay heed to and record what is known about "native knowledge and manipulation of narcotic plants". The strides of modernization, urbanization, and cultural change will soon wipe out such knowledge. Anthropologists have both the responsibility and the opportunity to preserve and use this information.

BIBLIOGRAPHY

Schultes, Richard Evans, 1973, Tropical American hallucinogens: where are we and where are we going? Cienciae Cultura, 25(6).

Sorenson, E. Richard, 1975, Visual evidence: an emerging force in visual anthropology. Occasional paper No. 1 of the National Anthropological Film Center.

E. RICHARD SORENSON 14

PHENOMENOLOGICAL INQUIRY IN ETHNOBOTANICAL STUDIES

Some years ago, in an article on the place of ethnobotany in the ethnopharmacological search for psychotomimetic drugs, Schultes (1967) commented on the importance of the term, 'ethnobotany'. This word, a concise conceptual framework for the understanding of man in relation to plant, came into stable academic usage more than a half century earlier, in a publication of the Smithsonian Institution's famous Bureau of American Ethnology (Robbins, Harrington & Freire-Marreco, 1916). Yet, to-date, it has received its most serious attention on its botanic side. Many ethnobotanical lists now exist; but there is much less information on the patterns of human behavior and culture which govern the use of the plants.

Recently greater understanding of ecology in relation to culturally patterned human activity has begun to focus increased interest on the ethno part of the dyadic concept presented by the word 'ethnobotany'. It is now becoming increasingly obvious that inquiry into culturally and socially patterned human behavior is important to the understanding of how and why people use behavior modifying plants and drugs and how such use might be related to the larger questions of cultural organization and human adaptation.

By definition, ethnobotanical plants are part of a way of life. They relate to human behavior in the sense that their use is part of a larger pattern of human activity characteristic of a way of life. Such patterning of human behavior in relation to plant use may be due to direct pharmacological or nutritive effect; but it can also be less direct, as, for example, the result of ceremonial requirements, a cultural belief, or even habitual work patterns associated with production or use of a plant. In all of these human behavior is patterned in relation to a plant, and this patterning is reflected in the chemistry of mind, emotion, and motor response of the community members. Because of this, it is as important to know about the plant-related behavior of a way of life as it is to know the chemical effect on neurophysiology. Both kinds of knowledge are necessary if we are to understand ethnobotany in relation to the human condition.

Some plants more dramatically affect human neurophysiological response. Variously called psychedelic, psychopharmacological, psychotomimetic, hallucinogenic, etc., they are often considered 'mind altering' by Western observers. However, it is important to recognize that the concept, mind altering is, at least in part, culturally defined. For example nonwestern peoples who used plants considered to be mind altering by Westerners may not necessarily think of them as such — any more than we Westerners usually think of coffee or cigarettes as mind altering (or, for that matter, a church service or a violent TV program), although, upon reflection, the effect of these upon mood and thought is obvious.

How to discover and demonstrate culturally specific patterns of human behavior in other cultures touches on cross-cultural methodological problems of an anthropological nature. There are culturally imposed cognitive barriers to be overcome. Ceremony, belief systems, values, enculturation, and communicative behavior must be examined.

We Westerners have the advantage of being members of a modern, expansive, technological, 'wired' culture. We are therefore generally aware of the fact of cultural difference in human behavior throughout the world. Yet our 'understanding' remains tied to our own cultural foundations. This obstructs fuller appreciation of expressions of human existence as it exists elsewhere. The fact of difference in human behavior is easy to recognize; but it is not so easy to go beyond a statement of this fact as a relatively simple description of observed difference phrased in the concepts provided by the culture of the observer. Deeper appreciation is blocked by cognitive barriers. As enculturated beings, we have little choice but to recognize, conceptualize, and articulate in terms of the precepts, categories, and values bequeathed by our own culture. Culturally specific patterns of awareness and understanding, including ideas of value, screen sensory input, interposing culturally limited modes of cognitive appreciation. Our understanding of what we encounter in other cultures is, thus, limited. It is perceived and communicated through the colored spectacles of our own culture according to a system of appreciation provided by it. Such barriers are not easy to transcend. With the possible exception of feral man, all human beings are products of culture. All live in way-stations of human cultural development.

Because of this it is useful, when examining other cultures, to obtain records which go beyond our own ability to understand and appreciate. This is particularly important when enquiring into such culturally 'colored' matters as human behavior and organization; but it is also important when dealing with aberrant or novel expressions of human organization within our own society. The undifferentiated raw information of the original event, provided by such records, can be richer and less skewed than a culture-bound verbal description of the same event.

Such phenomenological records also uniquely permit review of the original situation — a more fundamental kind of inquiry than that which can come from examination of what has been said about the same situation. Furthermore they permit this from a variety of perspectives. They also provide a means of analytically manipulating the data so as to facilitate perception extending beyond usual cultural or linquistic habit.

Synchronized sound-film records have become a most powerful contemporary tool for preserving the visual and aural data of passing events. Film possesses a unique ability to preserve a facsimile of the visual data of a scene by means of chemical changes in a light sensitive emulsion. Information is thus captured and preserved beyond that perceived by human observers. Although human beings select the subject to be photographed (and cameras make it possible), it is the film which actually takes the picture. Its light-sensitive emulsion literally takes light energy emanating from a scene to produce objective chemical changes which capture a permanent record of the pattern of light received. Because of this, film preserves data not only of what is 'seen' or 'selected' by the culturally programmed mind of the human observer, but also of what is not. Even when the camera is picked up, pointed, and turned off and on according to a cameraman's particular interests and concepts of appropriateness, it also gathers information interstitial to and beyond the cameraman's interests and awareness. Sound tape performs a similar function for aural information. Both produce, therefore, deeper records of events than is possible by relying on anyone's perceptive and descriptive abilities. As means of gathering data, they extend beyond the cultural barrier — beyond the individual's personal or cultural screen. This is not so possible when data collection relies on transcription of only that which has been perceived by a human observer.

These richer, more comprehensive records also permit re-examination of past events as often as needed and as new knowledge leads to new questions. They also permit careful examination and re-examination of details and relationships which may be too subtle, fleeting, or complex to be detected in the real time of daily life or under the demanding conditions usually provided by field work in other cultures.

In the study of human behavior, such records have proved valuable, in some cases critical. Mead and her colleagues clearly demonstrated the use of film to discover and demonstrate culturally specific patterns of behavior on Bali (Bateson & Mead, 1942; Mead & McGregor, 1951). Birdwhistell's discovery of subtle, visible components in communication from film records showed that film was indispensable for analysis of human interaction (Birdwhistell, 1952, 1970). Hall's demonstration of cultural difference in nonverbal human interaction made it quite clear that motion picture records were essential to study the culturally or socially derived patterns of human behavior (1959). Lomax (1972) has shown that patterns of human movement may relate to socioeconomic structure and the evolution of culture. In all of these studies critical discoveries were made from film records which contained information neither recognized nor appreciated at the time of filming. The film records used permitted extension of awareness and understanding beyond usual habits of perception.

My own first attempts to use film to discover and demonstrate culturally specific patterns of behavior were focused on child rearing and socialization practices in primitive cultures (Sorenson & Gajdusek, 1966; Sorenson 1968b, 1971, 1976). From these studies the theory and methods of the research film emerged (1967, 1968a, 1971, 1973, 1974, 1975a, b, 1976). In these studies cameras provided a record facilitating discovery of culturally specific patterns of human behavior and its organization. The still camera was used to capture visible, static data of a scene such as items, positions, and accountrements, including those not necessarily noticed at the time of filming. The motion picture camera captured otherwise unobtainable patterns and subtelties of process and development in human behavior and social interaction not recognizable at the time of filming, including the unexamined contextual visible situation in which the subject selected for filming was immersed.

Several years ago, while participating in a preliminary study of microevolution on Tongariki Island in the New Hebrides, I was in a position to make only personal observations. The social ramifications of kava use there were unfamiliar and complex, ecompassing nonverbal interaction as well as cultural patterning of social behavior. Because I was unable to conduct a research film study of these relationships, I could only note that the use of kava on Tongariki appeared to differ from the better known pattern of use in Polynesia. Preparation was different: rather than squeezing the soaked dried root to obtain the beverage, fresh root was chewed and spit into a container for a half hour 'curing' before drinking. This difference was relatively easy to note. More difficult to examine, however, were the differences in social context: Not so much the focus of convivial interaction with guests, as in Polynesia, kava use on Tongariki was typically a nightly gathering of several close associates who imbibed deeply to achieve a quiet, reflective state of intoxication. Potency of the drink not only was greater, but use was more intense. Community structure seemed to be related to its use; and even child behavior and fertility appeared to be affected.

To have inquired more deeply into kava use in relation to cultural values, social structure, and patterns of human behavior would have required records from which details of nonverbal interaction and social relationships could be abstracted and examined. Had I been able to prepare a research film record, I could have plotted and examined patterns of human interaction and social organization against the larger cultural framework, enabling me then to inquire into the dynamics of structured human contact, emotion, and communicative behavior in relation to kava use.

With Tongariki and its lesson on the loss of valuable data due to the absence of film records behind me, my more recent attempts to study Huichol Indian child behavior and human development in relation to peyote use focus on obtaining the filmed materials which are essential for research.

Although still in its first phase, this study has already begun to yield preliminary results: some aspects of Huichol early child handling are more similar to those of the Fore of New Guinea than they are to those of our own Western culture. For instance, young children generally participate in what is going on about them according to their own interests and attraction; there is little restrictive or directive supervision; competitive rivalry does not become part of the growing child's way of life, and play is usually gentle enough to permit participation of children of different ages. There seems to be little interest in forcing a child against his will, embarrassing him, or demonstrating superiority over him, either by adults or by agemates. Bullying is rare. Considerable leeway is given to the unsophisticated inclinations of young children; they are rarely coerced or disciplined as they attempt to join in or watch daily activities or ceremonial events in their own childlike ways. These New Guinea-like elements were easy to identify on the basis of my recently completed study of patterns of child handling and behavior of the Fore people (1968b, 1971, 1976, n.d.). Demarcation of the more specifically Huichol characteristics awaits more detailed analysis of films yet to be obtained. Studies to show how these New Guinea-like elements may relate to peyote use or how they may act to help sustain the formal or ceremonial structure of the Huichol society are underway.

Initial observations indicate that Huichol children are rather casually introduced to subhallucinatory doses of peyote at an early age, both in ceremonial settings and occasionally during daily life around the house where it may be ground into the tortilla batter. Hallucinatory doses, sometimes very large ones, are more typically provided in deeply religious or initiation ceremonies. Its ceremonial use, particularly its religious use, may also act somewhat to curtail birthrate.

We are adopting a combination of three kinds of sampling strategies (see Sorenson & Jablonko, 1975):

1. opportunistic sampling;
2. programmed sampling; and
3. digressive search

These rely on:

1. seizing the opportunity we 'see';
2. taking advantage of the collective knowledge of our society or culture as a sampling guide; and
3. looking into the unknown.

All of these strategies take advantage of the unique ability of film emulsions to record both recognized and unrecognized, appreciated and unappreciated visual information. They parallel three basic elements of scientific inquiry:

1. the significance-recognizing capability of the human mind;
2. an existing accepted, rationalized body of knowledge; and
3. the desire to learn.

Each strategy has its own advantages and disadvantages. Each skews the sample in different ways; but, in concert, they begin to balance one another so as to increase the informative potential of the visual records.

Employing these three filming strategies, the study of child behavior and human development among the Huichol, in relation to peyote use, is divided into three phases:

Phase 1. Initial field work and collection of data as research films.
Phase 2. Analysis of research film records to discover culturally specific patterns of activity, behavior, child handling, and socialization.
Phase 3. New fieldwork to test the hypotheses formed.

Phase 1 — We are still in Phase 1. Opportunistic sampling and digressive search are being emphasized. At this point in our study we still know little about Huichol behavioral diversity as it may exist across the Huichol Sierra. We are sampling events which are not yet understood, and we do this as we become aware of their occurrence. Because the film record constitutes our basic field data, careful detailed records are being kept of date, place event, and circumstances of filming. Dr Muller, the cinematographer in the project, has now established sustained residence in San Andrés and his photographic activities no longer present so much of an intrusion on a way of life where cameras hitherto were not welcome. Initially, ceremonial activities were the main subject of filming; but now that familiar relations with a number of Huichol families have been established, attention is being given to naturally occurring social interaction as evidenced by daily household activities. Over the next few years we hope to introduce the camera to a broader geographical area and to be able to film without disturbance in an increased variety of social, family, and ceremonial settings.

Phase 2 — Although there has been some preliminary review and analysis of the film records obtained to date, Phase 2 has not yet begun. This phase is to be the systematic review and analysis of the body of research films obtained. All films will be searched for episodes of recurrent patterns of human behavioral response. Considerable attention will be devoted to enculturation. Infant handling, nursing, physical interaction, affectionate expression, deference patterns, approaches to learning and instruction, and the manner of expression of basic emotion in various social contexts will all be examined.

Phase 3 — This will be a return to the field to check the validity of the deductions made from the research filmed record, and to follow-up ideas or leads which may have developed during the course of film analysis. In particular we will look into situations and classes of events which are underrepresented in the research film record, from which preliminary but not conclusive findings will have been drawn. We will also gather new filmed material to fill in areas of cultural behavior not well sampled previously.

The visual resources obtained in the course of this study will remain available as a permanent scholarly resource in the National Research Film Collection of the Smithsonian Institution with three major purposes:

1. to support findings already made, much as museum collections permit reexamination of artifacts and specimens with reference to conclusions which have been drawn from their previous study;
2. to provide basic data permitting extension of early findings; and
3. to facilitate new kinds of inquiry not previously considered.

These advantages make it possible to bring studies of human behavior and social organization, especially those of changing and vanishing ways of life, more deeply into the realm of phenomenological inquiry and validative analysis and, thus, more firmly into the realm of science. They provide a means by which we may more deeply and critically examine the ethno part of ethnobotany.

BIBLIOGRAPHY

Bateson, G. & Margaret Mead, 1942, Balinese character: a photographic analysis. New York Academy of Sciences, Special Publications. Vol. 2.

Birdwhistell, R. L., 1952, Introduction to kinesics. Louisville: University of Louisville Press.

Birdwhistell, R. L., 1970, Kinesics and context. New York: Ballantine.

Hall, Edward T., 1959, The silent language. New York: Doubleday.

Lilly, J. & E. Richard Sorenson, 1972, Children of the Toapuri. A research report film prepared for the National Institute of Mental Health.

Lomax, Alan & Norman Berkowitz, 1972, Evolutionary taxonomy of culture. Science, 177-228-239.

Mead, Margaret & F. C. MacGregor, 1951, Growth and culture: a photographic study of Balinese childhood. New York: Putnam.

Robbins, W. W., J. P. Harrington, & B. Freire-Marreco, 1916, Ethnobotany of the Tewa Indians. Bureau of American Ethnology Bulletin, No. 55.

Schultes, R. E., 1967, The place of ethnobotany in the ethnopharmacologic search for psychotomimetic drugs. In D. Efron (ed.), Ethnopharmacologic search for psychoactive drugs. Public Health Service Publication, No. 1645.

Sorenson, E. R., 1967, A research film program in the study of changing man: research filmed material as a foundation for continued study of nonrecurring human events. Current Anthropology, 8:443-469.

Sorenson, E. R., 1968a, The retrieval of data from changing culture: a strategy for developing research documents for continued study. Anthropological Quarterly, 41:177-186.

Sorenson, E. R., 1968b, Growing up as a Fore. Scientific report film presented at the Postgraduate Course in Pediatrics, Harvard Medical School.

Sorenson, E. R., 1971, The evolving Fore: a study of socialization and cultural change in the New Guinea Highlands. Ph.D. Dissertation, Stanford University.

Sorenson, E. R., 1973, Research filming and the study of culturally specific patterns of behavior. PIEF Newsletter of the American Anthropological Association, 4(3).

Sorenson, E. R., 1974, Anthropological film: a scientific and humanistic resource. Science, 186:1079-1085.

Sorenson, E. R., 1975a, Visual records, human knowledge and the future. In Paul Hockings (ed.), Principles of visual anthropology. The Hague: Mouton.

Sorenson, E. R., 1975b, To further phenomenological inquiry: the national Anthropological Film Center. Current Anthropology, 16:267-269.

Sorenson, E. R., n.d., Ecological disturbance and population distribution in the Fore region of New Guinea. In Willis E. Sibley (ed.), China to the Antipodes. The Hague: Mouton, in press.

Sorenson, E. R., 1976, The edge of the forest: land, childhood and change in a New Guinea protoagricultural society. Smithsonian Institution Press. In press

Sorenson, E. Richard & D. C. Gajdusek, 1966, The study of child behavior and development in primitive cultures. A research archive for ethnopediatric film investigations of styles in the patterning of the nervous system. Supplement to Pediatrics, 37(1), Part II.

Sorenson, E. Richard & Allison Jablonko, 1975, Research filming of naturally occurring phenomena: basic strategies. In Paul Hockings (ed.), Principles of visual anthropology. The Hague: Mouton.

Sorenson, E. Richard & Kal Muller, 1974, Huichol enculturation; a preliminary report. Paper delivered at the American Anthropological Association Annual Meeting, Mexico City, November, 1974.

active principles may be is, however, not known. There has been the suggestion that α- and β-asarone may be the responsible constituents, but this requires confirmation.

Mexico is one of the great centres for the use of hallucinogenic plants. There are a number still not thoroughly understood. In addition to visual hallucinogens, Mexican Indians employ plants that induce auditory hallucinations. One of these is *Heimia salicifolia* — 'sinicuichi' — which in recent years has been chemically studied and is well understood at the present time. Two of these, however, still await investigation. The Mixtec Indians of Oaxaca take puffballs to induce definite auditory hallucinations: two species are employed — *Lycoperdon marginatum* and *L. mixtecorum.* What chemical constituents in these fungi may be responsible we still do not know. The weedy bush known as 'zacatechichi' — *Calea zacatechichi* of the Compositae — has been valued as a medicine in Mexico for centuries, but it has recently been reported as an auditory hallucinogen amongst the Cholo Indians of Oaxaca, who call it 'thle-pela-kano' or 'leaf of god'. No hint of the hallucinogenic constituents is as yet available.

Sundry species of cactaceous plants are suspected to have hallucinogenic properties and to be employed in Mexico as narcotics, but further field work is needed for corroboration. Among the Tarahumares, *Epithelantha micromeris,* a species of Echinocactus and the giant columnar *Pachycereus pecten-aboriginum,* known as 'cawe', are used. Furthermore, throughout Mexico, the term 'peyote' is applied to a large assortment of species in the cactaceous genera *Ariocarpus, Astrophytum, Aztekium, Dolichothele, Mammillaria, Obregonia, Pelecyphora, Solisia* and *Strombocactus;* certain species of the Crassulaceae, Compositae and Leguminosae are similarly named. There are suspicions that some of these plants may be used for biodynamic properties, although it is not impossible that some are so called simply because of similarity to *Lophophora williamsii,* the true peyote, or to some ritual connection with it. It is here that chemical studies may yield interesting data. Mexico offers still more unsolved riddles. Is it true, for example, that the toxic seeds of *Rhynchosia* were used as intoxicants and that the beans of certain species of *Erythrina* were valued as hallucinogens? What is the psychoactive principle of the proven hallucinogen, *Salvia divinorum* of the Indians of Oaxaca? Further studies may elucidate these enigmas.

It is in South America, however, that new developments in our discovery of hallucinogens are taking place and where chemical studies are lagging behind ethnobotanical progress. In the Andes of Ecuador, the known poisonous plant *Coriaria thymifolia* of the Coriariaceae has been reportedly employed by highland Indians as a narcotic — the fruits are ingested to induce the sensation of soaring through the air. Although long known and feared as a cattle poison in the Andes, this plant still is enigmatic, and there is no evidence of what it may contain that has psychoactive properties. The fruits are said to be the part of the plant employed in Ecuador. A recent report of the hallucinogenic use of *Petunia violacea* in Ecuador is equally interesting. These Andean areas are the native home of our horticulturally important genus *Petunia.* No chemical studies have apparently been carried out on *Petunia* — but since the genus is solanaceous and is so well known in horticulture, it would appear that our lack of knowledge of its chemical constitution is hardly to be justified. A very recently discovered hallucinogen, a species of *Iochroma* of the Solanaceae, is still chemically enigmatic; used in the southern Colombian Andes, it may in former times have been more widely recognized as a narcotic. Another enigma is presented by the use for narcotic purposes of the fruits of the ericaceous *Pernettya furiens* in Andean Chile and of *P. parvifolia* in Ecuador. These fruits are known to be toxic, but the exact nature of the intoxicating constituent has not been determined.

There is still much to be learned about the total chemical composition of the San Pedro cactus of Ecuador, Peru and Bolivia: *Trichocereus pachanoi.* Its effects as a visual hallucinogen are due undoubtedly to the presence of mescaline, but is mescaline the only alkaloid in the tissues of this cactus? It seems rather doubtful that this phenylethylamine would occur without other allied bases. If they be present, what part do they play in the intoxication that San Pedro induces?

The southern Andes of Peru is the home of the tall *Lobelia tupa* of the Campanulaceae, locally known as 'tupa' or 'tabaco del diablo', the leaves of which are smoked as a narcotic. The plant contains lobeline and related alkaloids, but the exact nature of the intoxication has not yet been defined in the literature. This region is likewise the area where *Desfonainia spinosa* var. *hookeri* of the poorly understood family Desfontainiaceae is reputedly employed for its narcotic properties, but again little is known with certainty about the nature of its effects and the extent of its use. Still another narcotic not chemically investigated is the 'keule' plant — *Gomortega keule,* the only species in the anomalous ranalian family Gomortegaceae. Endemic to Chile, this plant has been reported to be employed for its effects on the central nervous system. Unless the essential oils in its fruit be the psychoactive principles, we have no idea what the intoxicating constituent may be. A further poorly understood narcotic of Chile is the 'arbol de los brujos' or 'latué' — the solanaceous *Latua pubiflora*. Discovered in the 1850's, its ethnobotany and chemistry began to be elucidated only during the 1970's.

In the central Amazon of Brazil, Indians formerly prepared a snuff presumably from the fruits of a tall jungle tree of the Fig Family: *Maquira sclerophyllia.* The use of this narcotic has apparently disappeared, and the snuff is no longer prepared. Little is known of the source tree, and nothing is known of the chemistry of the snuff. It may now be too late to learn anything, but an attempt should be made to salvage what is possibly remembered by elderly tribesmen.

In the Amazon — especially in the westernmost parts — the solanaceous genus *Brunfelsia* has long been the source of important medicines. Recently, it has been reported used as an hallucinogen — both alone and as an admixture with *Banisteriopsis*. To date, there is uncertainty as to what the psychoactive principle may be.

III

A recent collection of 'kinds' of 'caapi' or 'yajé' recognized by the Barasana Indians of the northwesternmost Amazon may be cited as an example of aboriginal perspicacity in recognizing subtle differences in plants that often the botanist cannot see: age forms, ecological variants or even, at times, atrophied monstrosities. A detailed study of these 'differences' would seem to be scientifically justified, especially so since natives often ascribe varying narcotic effects to these 'kinds'. The Barasanas, for example, recognize eight 'kinds', seven of which, on the basis of sterile voucher specimens, can be identified as *Banisteriopsis caapi* or *B. inebrians:* although the sterile voucher specimens are indistinguishable, they have very special Barasana names: 'kumua-basere kahi ma' (yajé for shamanizing); 'kahi-uko' (yajé-catalyst); 'wai-buku-lihoa-ma' (game animal head yajé); 'wenanduri-guda-hubea-ma' (the yajé that came inside the jurupari instrument); 'yaiya-suava-kahi-ma' (red jaguar yajé); 'wai-buhua-guda-hubea-ma' (yajé that came inside the jurupari fish swim bladder). One variety is said to cause the Indians to see people during the intoxication; another is reported to induce red colored visions. The eighth 'kind' is botanically recognizable as a distinct species, *Banisteriopsis rusbyana,* and is called 'mene-kahi-ma' or 'nyoko-buku-guda-hubea-ma' (the vine that came inside the jurupari instrument known as 'old star'). A similar situation has been discovered among the Tarahumares of Mexico who recognize several 'kinds' of 'hikuli' or peyote (*Lophophora williamsii*): what may be an old polycephalic specimen is called 'hikuli-walula-saeliami' (peyote of greatest authority). These Indians also class as other types of hikuli species in related cactus genera: *Ariocarpus, Epithelantha* and *Echinocactus*, some of which are known to be alkaloidal and, consequently, biodynamic.

IV

Many plant additives are now known to be employed in the elaboration of a number of South American narcotic preparations. Some of these additives may be inert, but some have been shown themselves to contain active principles or other constituents capable of altering the activity of the main narcotic.

Perhaps the greatest number of plant additives pertains to the preparation of the drink known variously as 'ayahuasca', 'caapi', 'natema', 'pinde' or 'yajé', employed primarily in the western Amazon and prepared basically from the bark of the malpighiaceous *Banisteriopsis caapi* or *B. inebrians.* The two most significant additives are the leaves of the rubiaceous *Psychotria viridis* and the leaves of another species of *Banisteriopsis, B. rusbyana.* These are never added to the drink together, although both contain the same hallucinogenic N, N-dinethyl-tryptamine. The interesting observation is that the drink itself is hallucinogenic because of its content of three β-carboline alkaloids, and that the natives add the leaves to lengthen and strengthen the intoxication. The tryptamines are not active when taken orally, unless they are in the presence of a monoamine oxidase inhibitor; the inhibitor is provided by the β-carboline alkaloids. Several solanaceous additives are used. Tobacco — *Nicotiana tabacum* — is an occasion admixture in the caapi-drink, especially in the Rio Negro basin. In the westernmost Amazon of Colombia, the toxic *Datura suaveolens* is frequently so used. Several species of *Brunfelsia,* although also employed alone as narcotics, enter into the list of additives to 'ayahuasca'. The rare *Juanulloa ochracea* is called 'ayahuasca' in the Colombian Putumayo, a most unusual name which suggests the possibility that it may be employed together with *Banisteriopsis,* especially since the toxic alkaloid parquine has been reported from the genus *Juanulloa.* Similarly, in the same region, the name 'amarón borrachero' for the pontederiaceous *Pontederia cordata* suggests a role as a biodynamic agent. Throughout the Amazon, other additives to the *Banisteriopsis*-drink, none of them chemically investigated for possible active principles, include: the maranthaceous *Calathea veitchiana;* the amaranthaceous *Alternanthera lehmannii;* and *Iresine* sp.; several ferns, *Lygodium venustum* and *Lomariopsis japurensis;* the loranthaceous *Phrygilanthus eugenioides;* the labiate *Ocimum micranthum;* a cyperaceous species; a species of Clusia of the Guttiferae; species of the cactaceous genera *Opuntia* and *Epiphyllum*; and the apocynaceous *Malouetia tamaquarina* and what appears to be a species of *Tabernoemontana.* In several parts of the Amazon, plant additives, which it has not been possible yet to identify, have been reported.

'Epena' or 'nyakwana' is the snuff made from the bark-resin of several species of the myristicaceous *Virola*, especially *V. theiodora,* in the upper Orinoco and the northwest Amazon. The Waiká Indians of the Rio Negro basin of Brazil often add to the potent Virola snuff — which owes its activity to a high concentration of tryptamines — a powder of the dried, crushed leaves of the aromatic *Justicia pectoralis* var. *stenophyllia* of the Acanthaceae. The natives assert that they add the *Justicia* primarily to give the *Virola*-snuff a pleasant fragrance, but these people on occasion prepare a snuff only from the *Justicia* leaves. It has not been possible to find tryptamines in *Justicia*, although at one time it was suspected that they were present.

V

Then: there are those hallucinogens still to be botanically identified. There may be a relatively large number, but we do know of a few from their vernacular names.

What, for example, is the 'talka' of the Tukanos of the northwest Amazon? Described as a shrub about three feet tall with three long leaves, uniformly green, 'like grass', and with three contorted roots about three and a half inches long, this plant has defied identification. Said not to have flowers, it yields a root which, when chewed either dried or fresh, is very bitter and induces hallucinations after two or three days. Among the Tukanos also there is a plant described as a leafy vine with straight stems and no nodules and thin, smooth bark; it has the Tukano name of 'vai-gahpi', meaning 'caapi of the fish'. Another unidentified Tukano hallucinogen has been called 'muhia-gahpi': said to be a small vine with small leaves.

The isolated Tanimukas of the Colombian Amazon employ an as yet unknown hallucinogen is preparing a drink for adolescent initiation rites boys. The drink is used much as is the *Banisteriopsis*-drink, but the natives distinguish between the two drugs. The bark of the root of an extensive lacticiferous forest liana, without the admixture of any other plant material, is subjected to long boiling to prepare the beverage. The liana may well be a member of the Apocynaceae, but no progress has as yet been made towards the botanical identification of the species.

And what plant might be the source of the magic 'woi' of the Yekwana Indians of the Venezuelan Orinoco region?

Similarly, it would be interesting to know the plant source of the clear, amber-colored and aromatic resin from a fresh tree that makes up an important part of the accoutrement of every medicine man of the Tukanoan group of Indians in the Colombian Vaupés. In especially difficult cases of diagnosis of disease, divination or other magic ritual, minute amounts of this resin in powdered form are sniffed. It is said to induce dizziness, and it may well also be hallucinogenic. Botanical and chemical identification of this resin will be of interest indeed, since the resin is locally called 'paricá', the same term applied to several powerfully psychoactive snuffs in the same region.

And what could we guess is the 'marirí' of the Mojo Indians of eastern Bolivia? "Whenever ... [the medicine men] had to interview the spirits, they drank a concoction prepared from a plant called 'marirí', similar to our verbena, which caused for twenty-four hours a general condition of excitement characterized by insomnia and pain". According to reports, the medicine-man tried to avoid drinking marirí "whenever he could operate without the narcotic" — an indication certainly of its great potency or toxicity.

The report of the Makusi use of "peppers as a stimulant and excitant" in British Guiana is provocative. Although identified as *Capsicum*, is it

Above
Tukanoan Indian with stems of three 'kinds' of 'Caapi' employed in the preparation of an hallucinogenic drink on the Río Vaupés of Amazonian Colombia. Photograph: Reichel-Dolmatoff

Right
Waiká Indian preparing leaves of *Justicia pectoralis* var *stenophylla* for powdering and mixing with snuff prepared from *Virola*-resin. Rio Maturacá, Estado do Amazonas, Brazil. Photograph: Schultes

Top
Coriaria thymifolia of the Andean highlands. A group of Ecuadorian Indians reportedly eat the fruit for the sensation of levitation. The active principle has not been definitely established. Photograph: Schultes

Bottom
Lycoperdon marginatum and *L.mixtecorum*. Mexican puffballs reported to be hallucinogenic but the active chemical constituents of which are still unknown.

Calea zacatechichi, an auditory hallucinogen used by the Chontal Indians of Mexico, the active principle of which is not yet known.

Justicia pectoralis var. *stenophylla*. The dried, powdered leaves are sometimes added to snuff prepared from the hallucinogenic *Virola*-resin among the Waiká Indians of Amazonian Brazil. It is also on occasion used alone to elaborate a psychoactive snuff. The active principle has not been definitely established.

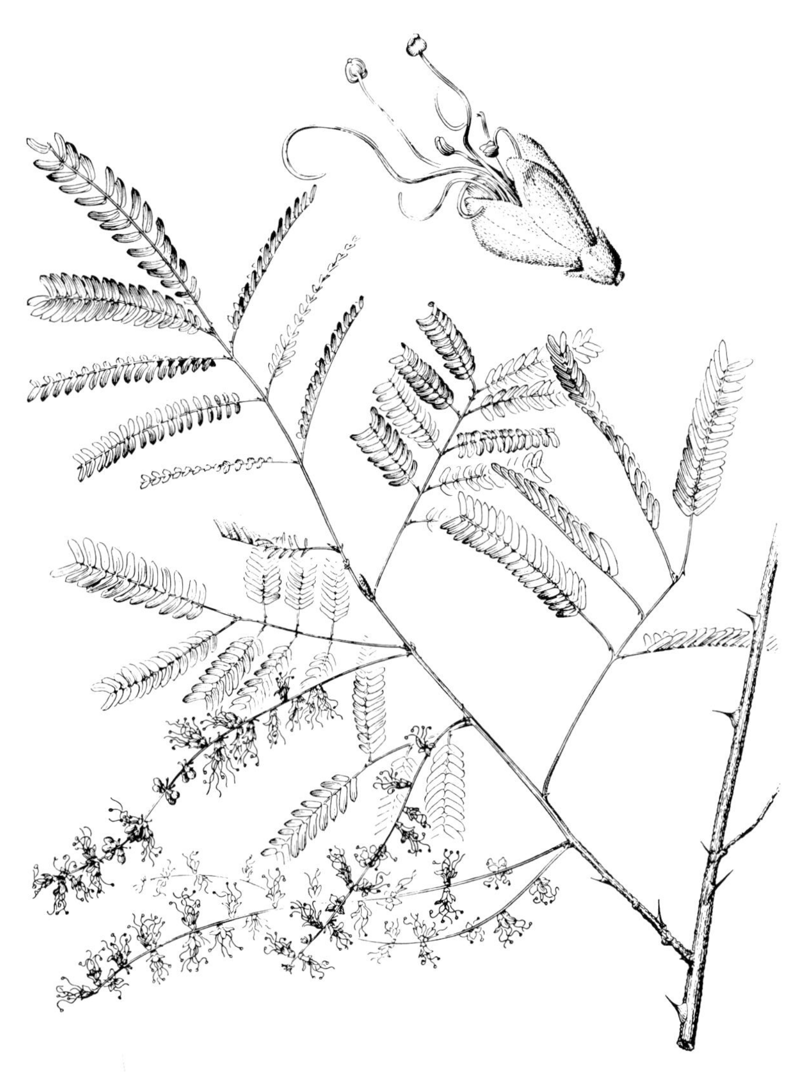

Mimosa hostilis, the root of which is the source of the hallucinogenic 'vinho de jurema' in Pernambuco, Brazil.

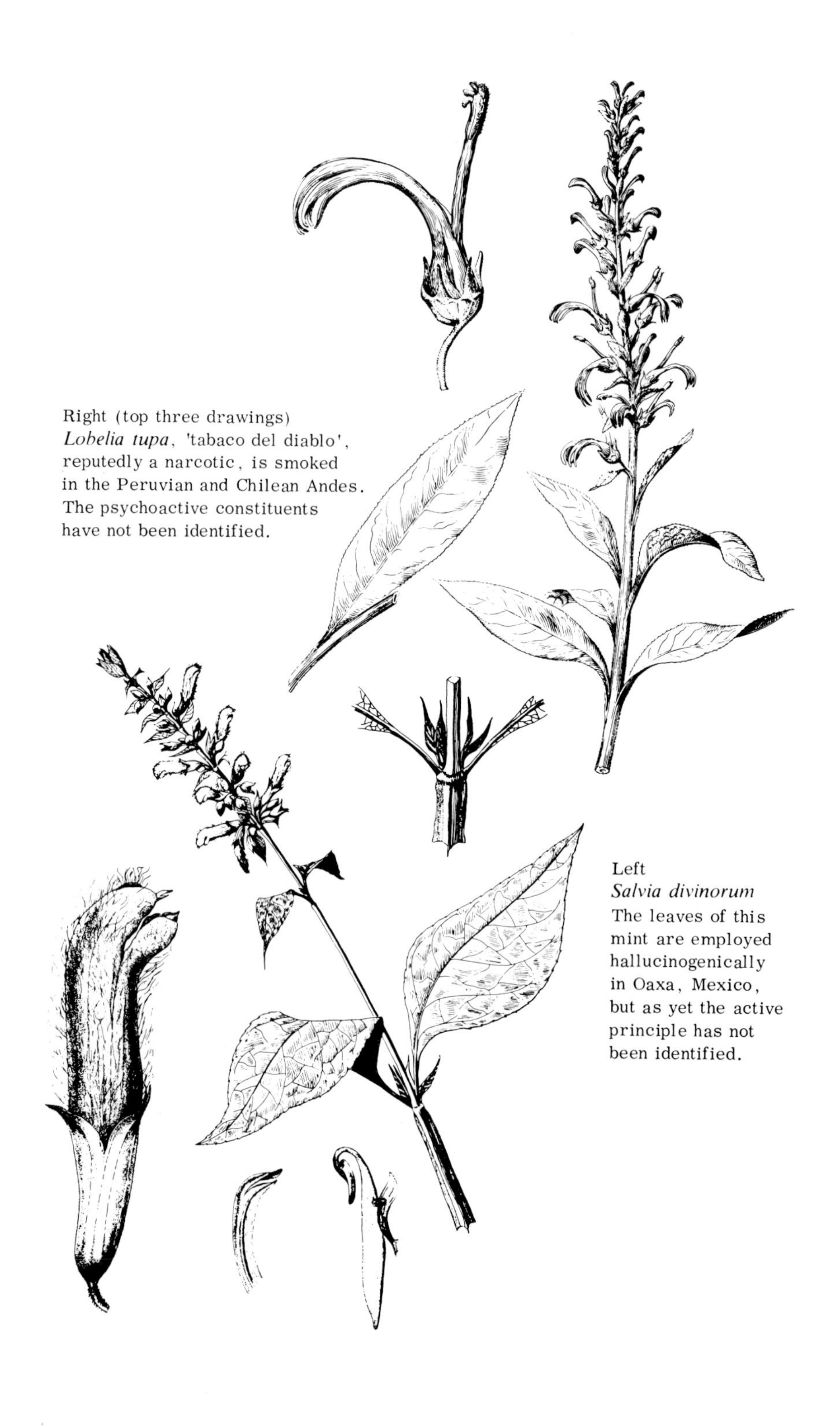

Right (top three drawings)
Lobelia tupa, 'tabaco del diablo', reputedly a narcotic, is smoked in the Peruvian and Chilean Andes. The psychoactive constituents have not been identified.

Left
Salvia divinorum
The leaves of this mint are employed hallucinogenically in Oaxa, Mexico, but as yet the active principle has not been identified.

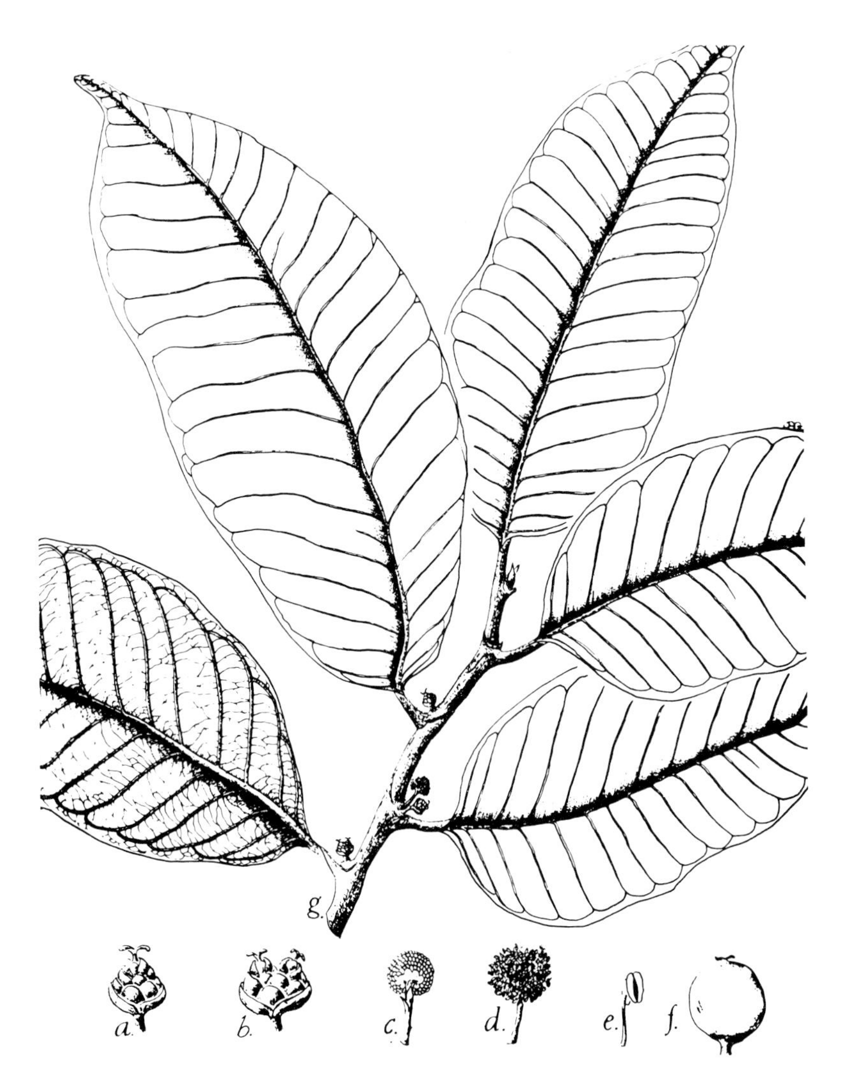

Maquira sclerophylla , reputedly the source of the hallucinogenic snuff 'rapé dos indios' of the central Amazon. Chemical studies are still lacking.

Tetrapteris methystica, a malpighiaceous liana of the Brazilian Amazon employed in the preparation of an hallucinogenic drink, 'caapi'. The psychoactive constituents, although probably β-carbolines, have not been established.

true that the plant involved is referrible to this genus or is it perhaps a member of some other highly pungent and irritating genus?

Although the hallucinogenic use of mushrooms has been well established in Mexico, modern investigations have uncovered no fungi taken as narcotics in South America. In the late 17th and early 18th centuries, however, missionaries found the Yurimaguas of the western Amazon drinking a potent intoxicating beverage prepared from a 'tree-fungus'. These Indians, it was reported, "... mix mushrooms that grow on fallen trees with a kind of reddish film that is found usually attached to rotting trunks. This film is very hot to the taste. No person who drinks this brew fails to fall under its effects after three draughts of it, since it is so strong, or more correctly, so toxic." It has been tentatively suggested that the mushroom might be the hallucinogenic *Psilocybe yungensis,* which has been collected in the region, but no evidence of its use in the Amazon has been found. Such a culture trait would seem little likely to disappear spontaneously without leaving some traces. And still one may wonder what the 'reddish film' said to be mixed with the mushroom might have been.

VI

That there remains much to be solved among the mysteries of the New World narcotics is probably quite clear. We have pointed out some of the obvious avenues that deserve future ethnobotanical attention. How many problems still unrecognized yet remain to be solved? The reasons for their solution are both academic and practical, but there is an urgency in the need for immediate attention. Aboriginal cultures almost everywhere in the world are threatened with extinction — together with the folk knowledge which forms a part of the cultures. The ever increasing inroads of modern civilization and technology are frighteningly fast, bringing the death knell to aboriginal knowledge and beliefs in totality and irreversible finality. We should pay heed to native knowledge and manipulation of narcotic plants before this valuable asset is forever snuffed from human consciousness with the demise of aboriginal cultures.

BIOGRAPHICAL NOTES ON THE CONTRIBUTORS

Michael Agar (1945-), born in Evanston, Illinois, received his B.A. in anthropology from Stanford University in 1967, and his Ph.D. from the University of California, Berkeley, in 1971. His field experience includes work among South Indian villagers, Austrian peasants, and urban American heroin addicts in Honolulu, San Francisco, New York, and Lexington, Kentucky. His publications include 'Ripping and running' (1973), 'Cognition and ethnography' (1974), and several articles on cognitive anthropology and the ethnography of urban heroin addicts. He is currently an associate professor in the Department of Anthropology, University of Houston.

Erika Bourguignon (1924-) was born in Vienna, Austria. She received the B.A. from Queens College, New York (1945) and the Ph.D. degree in Anthropology from Northwestern University (1951). She taught at The Ohio State University since 1949 where she is currently Professor of Anthropology. She is editor and co-author of 'Religion, altered states of consciousness, and social change' (1973), and co-author, with Lenora Greenbaum of 'Diversity and homogeneity in world societies' (1973). Her other publications include 'Culture and the varieties of consciousness' (1974) and 'Possession' (1975).

Katherine A. Carlson (1946-) was born in Modesto, California. She studied at the University of California, Berkeley, where she received her B.A. (1969) degree, and at the University of Hawaii, where she received her M.A. (1972) degree and her Ph.D. (1975). She is currently an instructor in anthropology at the University of Hawaii College of Continuing Education and at Chaminade College of Honolulu. Her other publications include 'Heroin, hassle, and treatment: The importance of perceptual differences', in Addictive Diseases, in press.

William E. Carter (1927-) was born in Dayton, Ohio. He studied at Muskingum College, where he received his B.A. (1949), Boston University, where he received an S.T.B. (1955), and Columbia University where he received an M.A. (1958) and Ph.D. (1963). He has taught at Brooklyn College, the University of Washington, and the University of Florida, where presently he is Professor of Anthropology and Director of the Center for Latin American Studies. Recent publications include 'Bolivia: a profile' (1971); 'New lands and old traditions' (1969); 'Trial marriage in the Andes?' (1976); 'Entering the world of the Aymara' (1972); and 'Chronic cannabis use in Costa Rica' (1976).

Patricia J. Cleckner (1946-) was born in Columbus, Ohio. She received her B.A. (1968) from Lake Forest College, and her M.A. (1971) and Ph.D. (1974) from Cornell University. She has done extensive field work among heroin addicts and Black drug users in New York City and Miami, having directly observed and participated in urban street life, therapeutic communities, and methadone programs. Her dissertation title was 'Dealing with Mister Jones: interaction between addicts and non-addicts in the street and in rehabilitation'. Articles by her have appeared in 'Journal of Psychedelic Drugs' and 'Addictive Diseases'. She is currently Director of Research at Up Front, a private drug information agency in Miami.

Marlene Dobkin de Rios (1939-) was born in New York City. She studied at Queens College of the City University of New York, where she received her B.A. (1959), at New York University, where she received her M.A. (1963), and the University of California, Riverside, where she received her Ph.D. (1972). She has taught anthropology at the City University of New York (City and Brooklyn Colleges), the University of Massachusetts, Boston, and California State University, Fullerton, where she is on leave from a position as Associate Professor. She is now an NIMH Post-Doctoral Research Fellow at the University of California, San Francisco medical school. where she is studying biofeedback technology and its adaptability to anthropological studies of altered states of consciousness. Specializing in the area of psychoactive plants and culture, she conducted fieldwork in the Peruvian Amazon among a transitional Indian population, who used plant hallucinogens in the treatment of psychological and emotional illness. She has published widely in scientific journals on her researches, including a book, 'Visionary vine: psychedelic healing in the Peruvian Amazon' (1972); in 1971, she prepared a monograph for the U.S. National Commission on Marihuana and Drug Abuse, entitled 'The non-western use of hallucinogenic agents'. A second book, based on this report, has just been completed, entitled 'The wilderness of mind: sacred plants in cross-cultural perspective'.

Brian M. du Toit (1935-) was born in Bloemfontein, South Africa. He studied at the University of Pretoria where he received his B.A. (1957) and M.A. (1961) degrees, and at the University of Oregon where he received his Ph.D. (1963). He was a lecturer in social anthropology at the University of Stellenbosch and at the University of Cape Town, and currently is Professor of Anthropology at the University of Florida. Recent publications include 'People of the Valley: life in an isolated Afrikaner community in South Africa' (1974); 'Akuna: a New Guinea village community' (1975); 'Migration and urbanization' (edited with Helen Safa, 1975); 'Configurations of cultural continuity' (1976); and 'Content and context in Zulu folk-narratives' (1976).

Wayne M. Harding (1947-) was born in Boston, Massachusetts. He was educated at Brandeis University (B.A. 1970) and the Harvard Graduate School of Education (Ed.M. 1971). He is currently Research Associate on the Social Controls of Nonmedical Drug Use Project at The Cambridge Hospital.

Thomas F. Johnston (1925-) was born in London. Following five years of schoolteaching in London, he studied at California State University, where he received his M.A. in music in 1968 and M.A. in anthropology in 1972. Following field work among the Tsonga of Southern Africa (1968-70), he completed a dissertation for the department of social anthropology at the University of the Witwatersrand, Johannesburg, earning the Ph.D. degree in 1972. He is currently Associate Professor at the University of Alaska, and holds a $60,000 three-year grant from the national Science Foundation, to research Alaskan Eskimo and Indian musical systems and their social foundations. Recent publications include 'Need for research on Botswana's sociocultural change' (Pan African Journal, 6(1), 1973) and about 75 articles on ethnomusicological folkloristic subjects.

John Bryan Page (1947-) is a Ph.D. candidate in anthropology at the University of Florida. He received his B.A. with high honors at the University and his M.A. at the University of North Carolina at Chapel Hill with the help of a Woodrow Wilson Foundation fellowship. After returning to the University of Florida because of his interest in the Latin American area, Page worked for one year as assistant to the director of the University's Urban and Regional Development Center under Carl Feiss. He began his two years of field work in Costa Rica as part of a transdisciplinary team assigned with the study of long-term chronic marihuana use in San José. Page's principal interests are Latin American studies, urban studies, drug research, and culture and personality.

William L. Partridge (1944-) was born in Miami, Florida, USA and studied at the University of Florida where he received his B.A. (1966), M.A. (1969) and Ph.D. (1974) in anthropology. He is currently Assistant Professor of Anthropology at the University of Southern California in Los Angeles. His recent publications include 'The Hippie Ghetto: the natural history of a subculture' (1973) and 'Cannabis and cultural groups in a Colombian Municipio' in 'Cannabis and culture' (edited by Vera Rubin, 1975). With Solon T. Kimball he is co-author of the forthcoming 'The craft of community study'.

Mercedes C. Sandoval (1934-) was born in Santiago de Cuba and attended the Universidad de la Habana (Cuba). She holds a Ph.D. from the Universidad de Madrid (Spain). During the past two decades she has been involved in a variety of educational programs in the U.S. as well as in Cuba, Puerto Rico and Spain. Currently she is Professor of Anthropology and Cuban Culture and History at Miami Dade Community College as well as Adjunct Assistant Professor of Psychiatry at the University of Miami School of Medicine. Her research on drugs was mostly done in her capacity as Research Director of the Spanish Drug Rehabilitation and Research Center of the Dept. of Psychiatry of the University of Miami. Dr Sandoval's publications deal with Afrocuban concepts of disease and its treatment and various aspects of mental health services. Her book 'Historia de la Santeria' was recently published in Madrid.

Richard Evans Schultes (1915-) was born in Boston, Massachusetts, and educated at Harvard University. After completion of his Ph.D. he was Plant Explorer in South America for the U.S. Department of Agriculture between 1943-1954. He then returned to Harvard Botanical Museum of which he is currently Director and P.C. Mangelsdorf Professor of Natural Science. In 1969 Schultes was recognized for his work in the Amazon when the Columbian government presented him with the 'Orden de la Victoria Regina'. He has been editor of 'Botanical Museum Leaflets' (Harvard University) since 1957 and of 'Economic Botany' since 1962. Schultes is the author of more than two hundred botanical communications, articles and chapters as well as author (or co-author) of five books, the latest being: 'The botany and chemistry of hallucinogens' (jointly with A.F. Hofmann, 1973).

E. Richard Sorenson received his Ph.D. in Anthropology from Stanford University in 1971. Now Director of the National Anthropological Film Center, Smithsonian Institution, he is presently engaged in worldwide studies of human potential and organization in cultural isolates. He is author of several publications and films dealing with child behavior and socialization, the relationship of social organization to ecological and demographic change, cultural factors in the expression of basic emotion, nutrition in a primitive economy, and cultural change. Pioneering an extension of scientific methodology into the study of nonrecurring phenomena through use of research film records, he has outlined the basic dynamics of proto-agricultural society, a distinct cultural form falling between hunting-gathering and settled agriculture, and his work on child behavior and human development in cultural isolates suggests the importance of tactile communication developing before language in young children. He has done fieldwork in New Guinea, New Hebrides, Cook Islands, Borneo, Micronesia, Brazil, and Mexico.

Norman Earl Zinberg (1921-) was born in Harrisburg, Pennsylvania. He was educated at the University of Maryland (A.B. 1942) and the University of Maryland School of Medicine. He has taught at and been on the staff of various universities and hospitals in the USA and Great Britain. He was appointed Staff Psychiatrist at the Cambridge Hospital in 1973. He has authored and co-authored more than eighty papers many of which deal specifically with various forms of drug use.